The OXFORD Children's A to Z of
Geography

Dick Bateman

OXFORD UNIVERSITY PRESS

Oxford University Press, Walton Street, Oxford,
OX2 6DP

Oxford New York
Athens Auckland Bangkok Bombay
Calcutta Cape Town Dar es Salaam Delhi
Florence Hong Kong Istanbul Karachi
Kuala Lumpur Madras Madrid Melbourne
Mexico City Nairobi Paris Singapore
Taipei Tokyo Toronto

and associated companies in
Berlin Ibadan

Oxford is a trade mark of Oxford University Press

© Dick Bateman 1996

First published in 1996
10 9 8 7 6 5 4 3 2 1

ISBN 0 19 910354 2 (hardback)
ISBN 0 19 910086 1 (paperback)

A CIP catalogue record for this book is available from the
British Library

Printed in Italy by G. Canale & C. S.p.A. - Borgaro T.se - TURIN

Acknowledgements

Design and art direction: Vivienne Gordon

Picture research: Image Select (London)

Abbreviations: t = top; b = bottom; l = left;
r = right; c = centre; back = background

Photographs
The publishers would like to thank the following for
permission to reproduce the following photographs:

Allsport at Image Select: 26l, 39r, 44b, 45, 64r;

Environmental Picture Library: 43b, 49b;

Gamma Presse: 46b;

Hutchinson: 56b;

Images Colour Library: 4b, 5c, 6t, 8c, 8t, 13 back, 14b, 18t,
21b, 23b, 23/24b, 24t, 27t, 30b, 31t, 32c, 33b, 34t, 35l, 40l,
40t, 41t, 41l, 41r, 41b, 52b, 53t, 54b, 54r, 57b, 58r, 62l;

Image Select: 34b;

Jacana: 10c, 11b, 12b, 16t, 23t, 27l, 34t, 36t, 48l, 49t, 50b, 52t,
64b;

Planet Earth: 31b;

Popperfoto: 29b;

Rex Features: 7b, 9t, 14t, 14l, 15b, 17b, 18b, 19b, 28b, 34/35b,
37t, 47l, 61r, 64 back, 64b;

Science Photo Library: 30t, 36b;

Spectrum Picture Library: 50l, 51t, 53b, 56b, 56r, 58b, 58t, 60b,
61r, 62b, 63c, 63b

Illustrations and diagrams
Michael Eaton: 48;

Helen Parsley: 1b, 2bl, 9, 19c, 37c,
52cr and heading panel, 53c,
54tr, 55c, 55t, 56t and heading
panel, 57b, 58b, 60c b and
heading panel, 63tr;

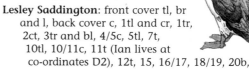

Lesley Saddington: front cover tl, br
and l, back cover c, 1tl and cr, 1tr,
2ct, 3tr and bl, 4/5c, 5tl, 7t,
10tl, 10/11c, 11t (Ian lives at
co-ordinates D2), 12t, 15, 16/17, 18/19, 20b,
22, 26, 28/29, 32, 38t, 40/41, 42/43, 45, 47,
49, 52l, 53c, 55b, 56cl and br, 57t, 58tl cr
and heading panel, 60tl and tr, 61c, 64 heading panel;

Martin Sanders: front cover tr, back cover c, 1tl, 3br, 16b, 21tl,
24c, 27, 28c, 29c, 31, 33, 37bl, 46, 51b, 52/53, 54/55, 56/57,
58/59, 60/61, 62/63, 64;

Peter Visscher: 2br, 3tl, 4l, 7b, 12b, 20-21t, 24t, 38-39b, 42b,
44, 51c, 52tr, 54cl cr bl and heading panel, 58cl, 59c, 61tr,
62t and heading panel, 63c

Many thanks for the contribution made by W.H. Ruthven of
the Mount Everest Foundation.

Dear Reader

Would you like to raft down the white-water rapids of the Grand Canyon, to watch elephant herds on the savanna of eastern Africa and to walk on the icy ridges of the Himalayan Mountains?

What's that? You don't know what the words **rapids**, **savanna** and **ridge** mean? Well, please read on to explore these and another 300 interesting words. You can find out about the plains, mountains, cities, countryside, continents and oceans which make up the geography of this wonderful planet of ours.

You will be able to use these words in school because they will help you to understand and describe places nearby and far away. At home you will know what has happened when you see on the news that a hurricane has hit the West Indies or that a volcano has erupted in eastern Asia. You can learn interesting facts like where to find snow on the equator and where icebergs as big as cities break away from cliffs of ice into the sea.

Here's wishing you many enjoyable hours of exploring the *A to Z of Geography* – and then, perhaps, the white-water rapids of the Grand Canyon...

Dick Bateman

◁ When there is a lot of acid in rain, it can kill trees, like this one.

A

acid rain

Acid rain is rain which has a lot of acid in it. A little bit of acid is found in all rain but it does not cause much harm. Acid rain is caused by certain gases which come from the chimneys of houses and **power stations**. These gases mix with rain to make acid rain which can cause a lot of harm. Acid rain kills trees, makes people cough and wears away buildings.

acre

An acre is a piece of land about the size of a small football pitch. Farmers used to measure their fields in acres (now many use **hectares**). A one-acre field was small; a ten-acre field was large.

Africa

Africa is the second largest **continent**.
See the **map on pages 56–57**.

agriculture

Agriculture is the way in which farmers use the soil for growing crops and raising animals. It is another word for **farming**.

alp

Swiss people call a high, grassy field on a mountain an 'alp'. The word is also used to name the Alps, a range of mountains in Europe.

altitude

The altitude of a place is how high it is 'above sea level' (a.s.l.). The altitude of the highest mountain, Mt Everest (or *Sagarmatha* in the language of Nepal), is 8848 metres a.s.l. See also **sea level**.

America

North America and South America are **continents**.
See the **maps on pages 52–55**.

annual

Something which is annual happens within a year. Annual rainfall is how much rain falls in a year. Annual can also mean something that happens once a year, like an annual holiday.

Antarctica

Antarctica is the frozen land around the **South Pole**. It is the coldest **continent** in the world.
See the **map on page 64**.

Antarctic Circle

The Antarctic Circle is a line which is drawn on maps of the area around the **South Pole**.
See the **map on page 64**.

▽ Agriculture in Japan: these farmers are harvesting a field of rice.

arable

Arable land is any land which is ploughed up by farmers to grow crops. Arable farmers grow crops such as potatoes, wheat, rice and peas. See also **pastoral**.

archipelago

An archipelago is a group of **islands** which are close together. The Greek islands in the Aegean Sea and the West Indies in the Caribbean Sea are archipelagos.

The arid landscape of the Atacama Desert in South America. In one part of this desert no rain fell for 400 years. △

area

An area is a piece of land. The area of a field or a city is how much land it takes up. To find the area of a place you multiply how long it is by how wide it is. If a playing field is 70 metres long by 36 metres wide, then it has an area of 2520 square metres.

arid

Arid means dry. The arid lands of the world are its deserts, where there is little rain and few plants can survive. You can tell a land is arid because what you see is mainly rock and sand. Some desert areas get no rain for years.

Arctic

The Arctic is the region around the **North Pole**. Most of the Arctic is sea, much of which is frozen.

 An ice-breaker ship pushes its way through the Arctic ice.

Arctic Circle

The Arctic Circle is a line which is drawn on maps of the area around the **North Pole**.

Asia

Asia is the biggest of all the **continents**. See the **map on pages 60–61**.

atlas

An atlas is a book of maps. You may have seen an atlas showing maps of countries of the world.

atmosphere

The atmosphere is the air above us. It is a thick layer of gases that surrounds the **Earth**. Above the atmosphere there is no more air, only space.

atoll

An atoll is a ring-shaped **island** made of **coral**. Many atolls have palm trees and sandy beaches. There are thousands of atolls in the Pacific Ocean.

Australia

Australia is the smallest of the world's continents. See the **map on pages 62–63**.

autumn

See **season**.

avalanche

An avalanche is a lot of snow and ice, and sometimes rock, falling very quickly down the side of a **mountain**. Each year many people die because they are caught in avalanches.

Admiral Beaufort (1774–1857) created a scale to describe the strength, or force, of the wind.

barometer

See **weather forecast**.

These few coral islands in Tahiti form an atoll.

bay

A bay is a curve in a **coast** where the sea washes up between two **headlands**. Many bays are small and are called coves. Some are huge. The Bay of Bengal between the countries of India and Myanmar is 1000 kilometres across.

beach

A beach is made of pebbles or sand. You find beaches where the sea or lake water meets the land.

bearing

A bearing is the angle of any direction from north. You can learn to take bearings with a **compass**.

Beaufort Scale

The Beaufort Scale is used for measuring the strength of the **wind**. It lists what will happen to natural things, such as trees, when the wind is blowing at different speeds. To use the scale you have to look at the effects of the wind.

Force 0: calm

Force 4: breeze

Force 8: gale

Force 12: hurricane

blizzard

If you were in a blizzard you would be fighting against a strong wind with lots of snow.

bog

A bog is wet, soft ground with moss and reeds. **Peat** forms in bogs. Animals, and sometimes people, can sink into the soft ground and may get trapped. See also **marsh, swamp.**

boundary

A boundary is the edge of any area. If you stand on the boundary of a field you will find a fence, a hedge, a wall or wire marking it off from the next field. Countries also have boundaries.

breeze

A breeze is a **wind** you can feel. A light breeze makes leaves move on trees.

bridge

People build bridges over rivers, roads and railways so that they can cross over safely.

bypass

A bypass is a road built around a town to keep traffic out of the town centre. Some people like bypasses, saying that they make towns safer, quieter and more pleasant for people. Other people are often against bypasses, saying that they spoil beautiful scenery and use up farmland.
See also **motorway.**

△ Many canals are not very wide and so narrow boats, or barges are used to transport goods from one place to another.

▽ These people are protesting about the building of bypasses through countryside.

C

canal

A canal is a large ditch which has been dug out, lined with clay and stone, and filled with water. Many canals are built as short cuts between rivers or seas. Canal barges and ships float on the water and carry **cargoes** from one place to another.

The Panama Canal in Central America connects the Pacific and Atlantic Oceans. The journey from New York to San Francisco by sea was 6000 kilometres longer before the Panama Canal was built.

canyon

'Canyon' is the Spanish word for a steep-sided **valley**. Canyons are found in dry parts of the world where powerful **rivers** have cut into the rock for many years.

The world's most famous canyon is the Grand Canyon in south-western USA. It has been formed by the Colorado River cutting into the Colorado **Plateau** over millions of years. It is 1.5 kilometres deep in places. With its red and yellow rock, the Grand Canyon is a spectacular sight.

cape

A cape is the end of a country or continent, sticking out into the sea. For example, the Cape of Good Hope is on the tip of South Africa.

△ The steep sides and grooves of this canyon are made by strong rivers, wind and rain.

capital

There are two meanings for this word. The capital of a **country** is the city where people run the country. Capital is also the name of the money needed to start up a business.

cargo

A cargo is a load of **goods** being carried by a lorry, train, aeroplane or ship.

cartography

The science of drawing **maps** is called cartography.

cave

A cave is a hole made by a river or the sea in the rock of a cliff or hill. Caves are big enough for you to get into. A large cave is sometimes called a cavern. Rivers flowing underground make large cave systems, which cavers explore for fun. See also **erosion**.

cereal

See **crop**.

This painting is on the wall of a cave in Lascaux, France. It was painted by people living in the cave in ancient times, perhaps 17 000 years ago.

chalk

See **limestone, rock**.

channel

A channel is a narrow stretch of water. It may be part of a river or a narrow stretch of sea linking two bigger seas or oceans. The English Channel links the North Sea with the Atlantic Ocean and is an important **route** taken by many ships.

city

A city is a big town, a place with lots of people. Mexico City is the biggest city in the world. It has 15 million people, making it as big as New York and London put together.

clay

Clay is a very useful soft **rock**. We use it to make many things, including crockery, like cups and plates, and bricks for building houses.

cliff

A cliff is a steep face of rock. We see cliffs at the seaside and in the mountains. Sometimes they are hundreds of metres high. Seaside cliffs often provide safe nesting places for large numbers of birds.

Cliffs like these, in Australia, are slowly being worn away by the moving sea water and waves.

climate

The climate of a place is its **weather** over the whole year. Different parts of the world have different climates.

Near the North and South Poles the climate is very cold, and ice covers much of the sea and land. In the Tropics the climate can be hot and wet, or hot and dry. Rain forests grow in a hot, wet climate. Deserts are found in hot, dry climates. Parts of North America and Europe, including the UK, have a temperate climate. This means it is neither very hot nor very cold.

coast

The coast is where the sea meets the land. A stretch of coast is marked on a map with a line. This line is called a coastline.

Two very different climates: in the Sahara Desert, Africa, the daytime temperature can be 58°C; the two men above are in Siberia, Asia, where it can be as cold as −60°C.

clouds

Clouds are patches of water droplets which we see floating in the sky. Inside they look just like **fog** but from the ground we often see them in different shapes and colours. Clouds that look like wisps of white hair are called cirrus clouds.

communications

Communications are the ways in which people travel or pass their ideas from one place to another. For example, you can travel on a motorway to talk to someone, or you can talk to them over the telephone. Communications include railways, roads, waterways, telephones, televisions and newspapers.

coal

Coal is a black, shiny **rock**. It has formed from plants that lived millions of years ago. We can burn coal to give us heat and light.

A coal-mine is a place where people dig coal out of the ground. Another word for coal-mine is colliery.

commuter

A commuter is someone who travels quite a long way to work. Commuters usually go by car or train. Most commuters go to the big cities. For example, thousands of people travel into Rome each morning, do their work, and then go home again in the evening.

Coal-mines are often hundreds of metres below the ground. Work in a coal-mine can be very dirty, and dangerous.

compass

A compass is a small box which holds a thin magnet. The red end of the magnet points to north. You can use a compass to find where north is. Then you can find where all the other directions are. This helps you find your way.

In a compass, north is the direction 0° (0 degrees). South is the direction 180° from north, east is 90° from north, and west is 270° from north. When you face north, south is behind you, east is to your right and west is to your left.

condensation

You may have noticed windows sometimes mist up with tiny water droplets. This is condensation. It happens when air is cooled, by a cold window for example. **Clouds**, **dew** and **fog** form in the same way, when air cools.

coniferous

Coniferous trees are trees with cones. Most have leaves shaped like needles. A pine tree is a coniferous tree.

 The giant redwood tree is a coniferous tree.

conservation

Conservation means looking after the plants and animals living around us. It is the protection and careful use of forests, rivers, seas, countryside, soils and other parts of your natural surroundings. Putting up nesting boxes for birds is one form of conservation. There are also organizations that raise money and run campaigns to protect and conserve areas like **rain forests** or rare animals like the tiger and the rhinoceros.

continents

The continents are the seven largest areas of land in the world. They are Africa, North America, South America, Antarctica, Asia, Oceania and Europe. See the **maps on pages 52–64**.

contour

Map-makers draw lines called contours on their maps. Contours join up places that are the same height and show how high places are above **sea level**. We say that sea level is the 0-metres contour. Imagine you are standing where the **X** is on the map below. You would be standing on the 100-metre contour. This means you are 100 metres above sea **level** – even if the sea is a long way away. See also **relief**.

150m

50m
100m

X

These contours show an island with a hill 150 metres above sea level. All coastlines are at 0 metres – this is sea level.

conurbation

A conurbation is a large built-up area caused by several towns growing into one another. The Tokyo-Yokohama conurbation in Japan is made up of places that grew up together. This conurbation has more people than any other built-up area in the world.

co-ordinates

Co-ordinates are numbers or letters telling you where a place is on a **map**. On the map shown, Sue lives at A5 and Bob lives at B4. A1 and B2 are co-ordinates. See also **grid**.

coral

Coral is a rock made from the shells of tiny animals living in warm sea. Coral often forms sharp ridges of rock which can tear holes in the bottom of ships which sail over them. See also **reef**.

On this map, who lives at the co-ordinates D2?

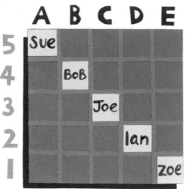

council

A council is a group of people who decide what is best for the people who live in their **city**, **county**, district or **town**.

country

A country is a land with its own government. It has its own flag and its own laws. The biggest country in the world is Russia. It is almost 8000 kilometres from west to east, and takes nearly a week to cross by train. The smallest country is the Vatican City in Rome. You can walk across it in one hour.

countryside

The countryside is the name for the land away from towns. It may have fields, grassland, woodlands, moors, lakes, hills and villages.

county

A county is an area inside a **country**. Each county has its own **council** to look after its people.

In the warm sea water where coral grows, there are also hundreds of colourful fish.

core

See **Earth**.

corrie

A corrie is a large bowl-shaped hollow. Corries are found on high mountains. Sometimes they have filled up with water to become lakes.

Corries were formed by ice eroding the rock, and would once have been the starting point of small **glaciers**.

150 m
100 m
50m
0 m

cove

A cove is a small **bay**.

The crater at the top of a volcano can be a hot, dangerous place.

crater

A crater is the hollow at the top of a **volcano**. Craters are also sometimes formed by a meteor from space that crashes on to the surface of the land. There are many craters on the surface of the Moon.

crevasse

A crevasse is a deep crack in a **glacier**. Crevasses can open and close as the glacier moves. They can also be hidden by layers of snow, so it can be dangerous to walk or climb across a glacier.

crop

A crop is a plant grown by farmers. Rice, wheat and potatoes are the main crops that the world's farmers grow. Rice and wheat are types of cereal crops. Cereal crops have been grown over centuries for their grain. Breakfast cereals are made from these plants.

A fodder crop is any crop grown so that it can then be fed to farm animals such as cows. Hay and straw are grown for animal fodder.

crust

See **Earth**.

cultivation

Cultivation is what farmers do when they plough the ground and grow crops.

Not all land can be cultivated – some is too steep, too wet or too dry for crops to be grown.

This farmer is cultivating the land. He is ploughing the land so that he can grow ① corn, ② rice and ③ wheat.

current

There are three kinds of current.

A river current is the fastest part of a river's flowing water.

An ocean current is moving ocean water that flows for thousands of kilometres. An example of a current is the Labrador Current which carries icebergs from Greenland southwards towards Newfoundland.

An air current is a stream of moving air. The 'Roaring Forties' are air currents. They blow from west to east all round the world between 40 degrees south (40° S) and 50° S.

cyclone

A cyclone is a huge storm which starts over warm seas. The people of the coasts of India and Bangladesh live in fear of the strong winds, high tides and flooding which cyclones bring. Cyclones are often called hurricanes or typhoons.

D

dairy

Dairy farming is where cows are reared for their milk. A dairy is where the milk is cleaned and put into bottles for us to drink.
See also **pastoral**.

dam

A dam is a wall built across a valley to hold back river water and form a lake. This lake is called a **reservoir**.

The Aswan High Dam across the River Nile in Egypt is 3.6 kilometres long and 111 metres high.

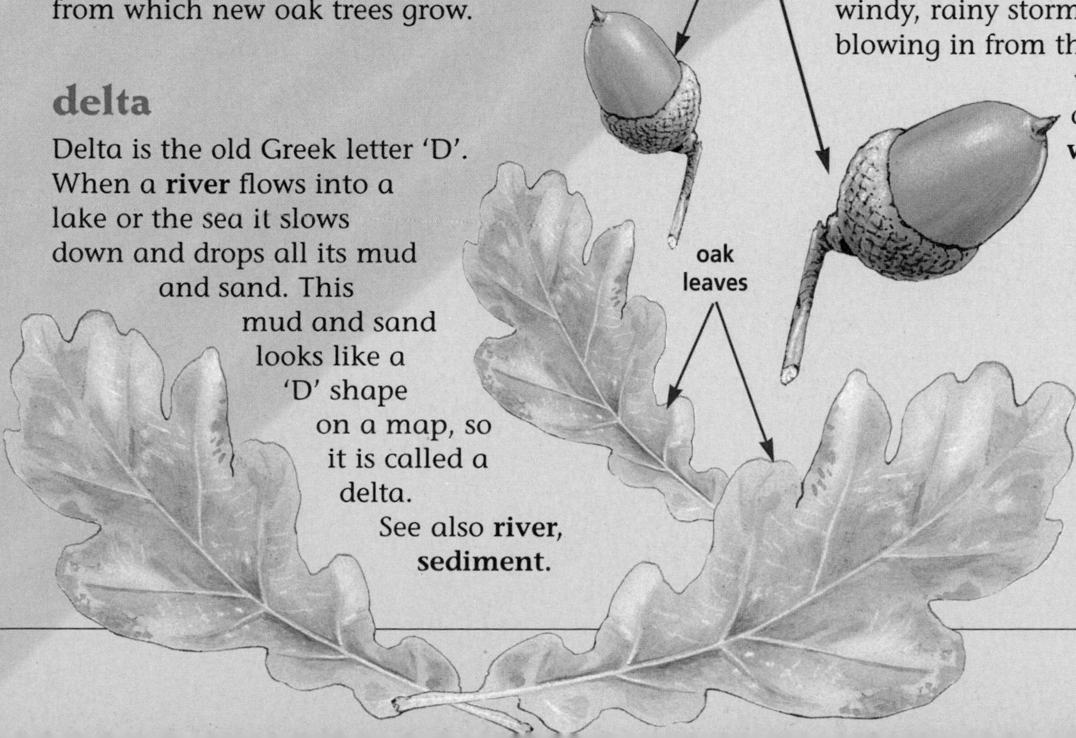

deciduous

Deciduous trees lose their leaves in winter, when there is not enough sunshine for them to make food. An oak tree is an example of a deciduous tree. Oak trees are known for their acorns, from which new oak trees grow.

delta

Delta is the old Greek letter 'D'. When a **river** flows into a lake or the sea it slows down and drops all its mud and sand. This mud and sand looks like a 'D' shape on a map, so it is called a delta.
See also **river**, **sediment**.

acorns from an oak tree

oak leaves

deposition

Deposition is the dropping of mud, sand and pebbles when moving water is slowed down. You can find mud or sand or pebbles deposited at the bottom of every stream, river, lake or sea.
See also **sediment**.

depression

Depression has two meanings in geography.

A depression in the earth is where the land is low.

The word 'depression' is also the name given to a windy, rainy storm blowing in from the sea, with warm and cold **weather fronts**.

The fleshy leaves of these desert plants help them to save water.

desert

A desert is a dry land with very few plants. One kind of desert plant is shown above. When you look at a desert you see lots of bare rock and sand.

13

In a developed country you will find wide roads and a large number of cars.

developed countries

Developed countries are rich countries where people work in many types of jobs. They have already built homes, roads, factories, schools and hospitals. See also **developing countries**.

Transport in a developing country, such as India, needs to be improved.

developing countries

Developing countries are poor countries where most people work on farms. The countries are developing as people improve their ways of farming and start new industries like tourism. This brings in more money to improve things like roads, water supplies and factories.

dew

Dew is the drops of water you find on the grass in the evenings and mornings, even though it has not rained. Dew is caused by **condensation** of water drops from the air at night.

directions

Directions show you the way you want to go. North, south, east and west are directions. To find north you can use a **compass**. A compass needle points to north. You can also find north by standing at midday with your back to the sun. North is then in front of you. This only works if you live north of the equator (for example, in Europe, Asia or North America). If you live south of the equator you must stand with the sun in front of you at midday. North is then also in front of you.

disaster

A disaster is an event which causes great hardship and suffering. It may be natural, as in an **earthquake**, or it may be the result of something people do, such as an aeroplane crash.
See also **war**.

distance

Distance is how far it is from one place to another. It may be a short distance across a room, or a long distance from one side of a country to another. The distance round the world at the equator is 40,076 kilometres.

dock

You may have seen ships or barges being loaded or unloaded at a dock. Docks are usually found in ports by the sea, but big rivers and canals also have them.

downstream

See **river**.

When there is a drought the ground can dry up and crack.

drought

A long period of dry weather is called a drought. Deserts often have no rain for several months, or even years.

dune

See **sand**.

Earth

Earth is the name of the planet we live on. It is a huge ball of rock, with water on its surface, and surrounded by air.

The outer edge of the Earth is called the crust or bedrock. It is the solid rock beneath the soil. We see it in cliffs and quarries. Under this is the mantle. Some of it is solid rock and some molten rock.

Deep down under the crust there is hot runny rock called magma. Where the crust has cracks in it, this runny rock flows out to the surface as lava and makes a **volcano**.

At the centre of the Earth is the core. The rocks at the core are 6000 kilometres beneath our feet, and are very hot.

Soil may sometimes be called earth.

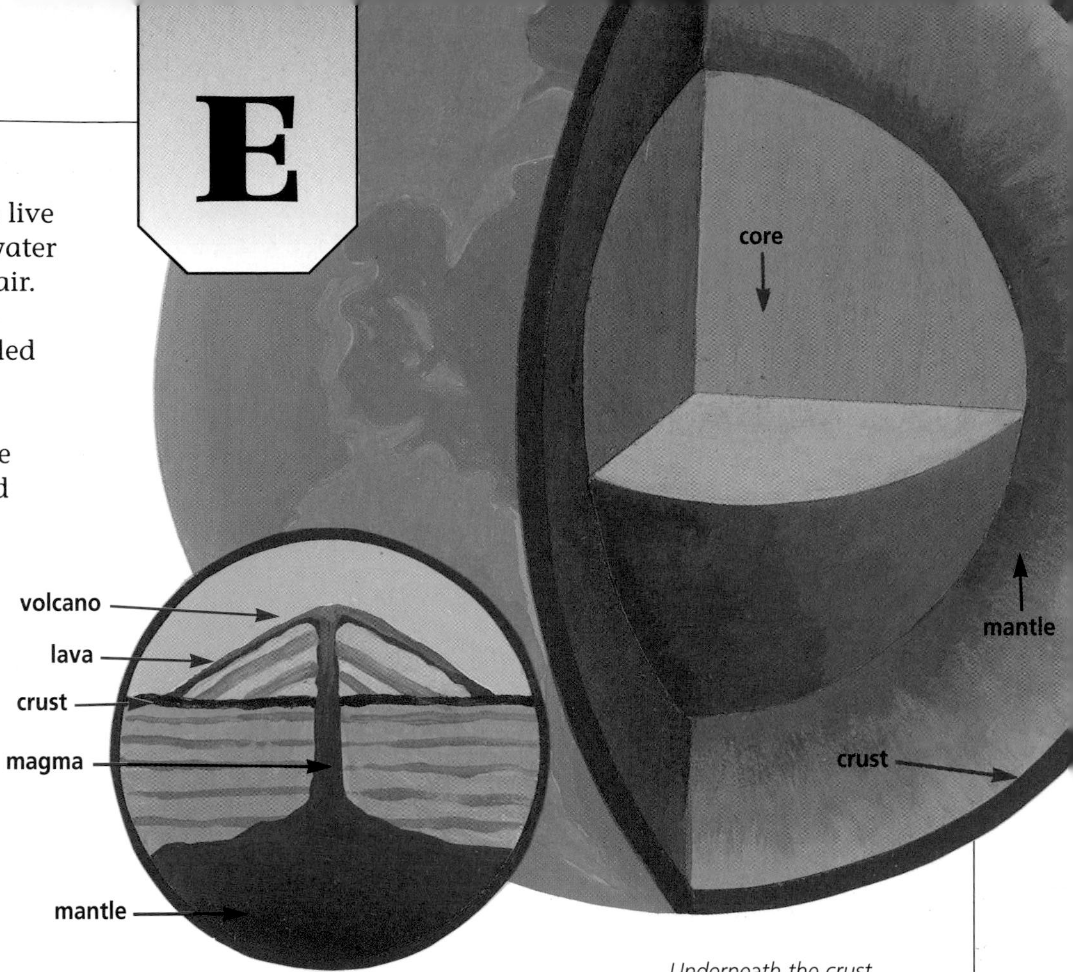

core

mantle

crust

volcano

lava

crust

magma

mantle

Lava is hot, runny rock which sometimes seeps out through a crack in the Earth's crust.

Underneath the crust, the inside of the Earth is very hot. At its core the temperature is over 5000° C.

earthquake

In an earthquake the land shakes and houses can fall down or huge cracks may appear in the ground. It happens when rocks in the Earth's crust move, deep under our feet. Earthquakes are more likely to happen in some places than others, and scientists are getting better at telling when and where an earthquake will hit.
See also **Richter Scale.**

When earthquakes happen near cities, they can cause a lot of damage to buildings. This picture shows buildings in San Francisco after an earthquake.

east

See **compass.**

ecology

Ecology is the study of the balance of nature between animals, plants and the places in which they live. People are animals and a part of this balance, too. Ecology helps people to understand that what they do affects the plants, animals, air, water and soil around them. Ecologists have shown us how acid rain affects forests.

economy

The economy of a country is all the making, buying and selling of **goods** and services that goes on in that country. A country with a strong economy sells more goods and services to other countries than it buys from them. This makes its people richer.

These fish are part of a large ecosystem. This ecosystem is the sea and all the amimanls, fish and plants that live there.

ecosystem

An ecosystem is a community of plants and animals living in a particular **environment**, or place. A garden pond with its pond weed, frogspawn, goldfish and dragonflies is an example of a small ecosystem.

environment

Our environment is what is all around us. Any changes to an environment affects the plants, animals and people living there. If people cut down trees in **rain forests**, the environment for rain forest animals will have gone and the animals will die.

The environment can be damaged by many things that people do. For example, driving cars causes **pollution** which damages the environment.

equator

The equator is a line drawn on **maps** to show places that are half-way between the North Pole and the South Pole. Places near the equator are hot all through the year. The equator passes through **South America**, **Africa** and the islands of south-eastern **Asia**.

equatorial forest

See **rain forest**.

erosion

Erosion is how the land is worn away by water, ice or wind. Erosion is changing the shape of our world all the time. It is also how bits of rock, sand, mud and dust are made smaller.

At the seaside, wave erosion breaks cliffs into sand. Rivers erode the land to make valleys, and in the rivers themselves erosion turns pebbles into sand and mud. High up in mountains the ice in glaciers also erodes the land.

In deserts, wind erosion blows sand at solid rock and cuts it into strange shapes.

eruption

An eruption is when the top of a **volcano** blows up. Bits of the volcano are blown high into the air and come down as ash on nearby areas.

North Pole

equator

South Pole

The equator passes through South America, Africa, and the islands of south-eastern Asia.

escarpment

An escarpment is high land which slopes very steeply one way and very gently the other way. If you were cycling you would have to push your bike up the steep slope, but then you could free-wheel down the gentle side.

estuary

An estuary is the wide mouth of a **river** where it meets the sea. The level of the water rises with each high **tide** and falls with the low tide.

At low tide estuaries have large areas of wet mud and shallow water. It is then that you will often see large flocks of wading birds feeding in the estuary. Their long legs allow them to wade in the water.

Eurasia

Eurasia is the name given to the land made up by parts of the continents of **Europe** and **Asia**. The huge country of Russia is partly in Europe and partly in Asia.

Europe

Europe is a **continent**. See the **map on pages 58–59**.

In a violent eruption red-hot rock, ash and steam shoot into the sky.

evaporation

After it has rained and the sun comes out, any puddles dry up and disappear. This is partly caused by evaporation. The heat of the sun makes the water turn into a kind of gas called water vapour. The water vapour mixes with the air, and the water in the puddle disappears.

evergreen

Evergreen trees keep their leaves all through the year. Most **coniferous** trees and trees in **tropical** forests are evergreen.

export

Exports are things made or grown by people in one country and then sold to people in another country. For example, coconuts are exported from islands like Fiji in the tropical Pacific Ocean. They are sold to countries like Germany, where coconuts do not grow.

These coconuts are being gathered by a man in Sri Lanka ready to be exported to another country.

F

factory

A factory is a building where people make things. Cars and televisions are made in factories. Some modern factories have few people working in them. Most of the work is done by robots, with a few workers making sure the robots are working properly.

△ *A car factory, where robotic arms are used for joining parts together.*

farming

Farming is how farmers grow **crops** and rear animals. The main types of farming are **arable** farming (growing crops) and **pastoral** farming (rearing animals). The main crops are cereals like rice, wheat and barley. The main animals are sheep, cows, chickens and pigs.

road

fault

road

Earth's crust

famine

A famine is when there is not enough food for the people of an area. Some famines are caused by lack of rain. Other famines are caused by wars. The lands on the edge of the Sahara Desert in Africa have had the worst famines in recent years.

farm

A farm is the land and buildings used for growing **crops** and looking after animals. In some parts of the world, farmers grow food for their families on farms as small as one hectare (100 metres long by 100 metres across). The world's largest farm is a cattle farm in **Australia**. It is over 1 600 000 hectares in area (160 kilometres long by 100 kilometres across).

◁ *An Ethiopian mother and son. Ethiopians have suffered many famines. (Ethiopia is in Africa).*

fault

A fault is a split in the **Earth's** crust. At most faults the rocks no longer move. At other faults the rocks do move, causing **earthquakes**.

△

Along this fault, one part of the Earth's crust is lower than the other.

fauna

Fauna is the name given to all the animals of an area.

fertilizer

Fertilizer is something which is put onto the **soil** by farmers to make **crops** grow well. Fertilizers can be natural, like manure, or chemicals, which are made by scientists.

field

A field is a piece of land with a boundary, like a fence or hedge. Fields are used for **farming** and sports.

fiord

Fiord comes from a Norwegian word. It means a long, narrow **valley** filled with sea water. Fiords are found on coasts with high mountains. Norway in northern Europe and New Zealand in the Pacific Ocean have fiord coasts.

fishing

Fishing is the job of catching and then selling large amounts of fish. Most fishing is done in the world's oceans, in cold waters.

The people of China catch more fish than those of any other country. They catch 13 million tonnes of fish each year. The total amount of fish caught everywhere in the world is 100 million tonnes each year.

fault

fjord

Fjord is the Norwegian spelling of fiord. See also **fiord**.

flood

Floods are where water spills onto the land. They happen when river, lake or sea water rises enough to flow over onto nearby lowland. You see most floods after long spells of heavy rain, or when the sea is pushed inland by **gales** and high **tides**.

The fishermen of Sri Lanka have a special way of fishing – they balance above the water on stilts.

flood plain

See **river**.

flora

Flora is the name given to all the plants of an area.

fog

Fog is a cloud at ground or **sea level**. It is caused when damp air is cooled. Fog is thicker than **mist**. You can see further than 100 metres through mist, but you cannot see things 100 metres away through fog.

forest

A forest is a very large area of trees. **Hardwood** forests are found near the equator. Big **softwood** forests are found further north.

Many forests are being cut down by people. For example, half the world's rain forests have been destroyed in this way. The trees are used for wood, and the cleared land is used for farming.

forestry

Forestry is the job of looking after **forests**. Foresters make sure that young trees grow properly, so that other trees can be cut down and sold. If one young tree is grown for every tree that is cut down, the forest will stay the same size, and all will be well.

In Bangladesh, heavy rains cause floods. The people try to carry on as normal.

fossil

Many dinosaur fossils have been found. The fossil bones in this skeleton are from the giant stegosaurus dinosaur.

Fossils are the shapes of plants and animals which are found in rocks. They tell us that plants and animals lived and died over 200 million years ago, when the rocks were just sand or mud. Over the years the sand and mud turned into rock but the shapes of the dead plants and animals have stayed clear enough for us to see.

freight

Freight is the **goods** carried from the people who sell them to the people who buy them. Most freight on land goes by lorry. Heavy loads go by train. Over long distances, light valuable goods go by aeroplane, but heavier goods go by ship because it is cheaper.
See also **cargo.**

frontier

A frontier is the boundary of a country. It is usually marked in some way so that you know you are leaving one country and going into another.

frost

Frost is made of ice crystals. You see it shining white on the ground or on wires and trees in winter, when everything is very cold.

Some of the tools used to study geology. A hard hat is important for safety.

geological hammer

safety helmet

an Abney level measures rock angles

gale

A gale is a strong wind. It breaks twigs off trees and is very hard to walk against.
See also **Beaufort Scale.**

geography

Geography is how you find out about the places and people in our world.

Human geography is about the human features of the world. It is about people, countries, towns and cities, and farming and industry.

Physical geography is about the natural features of the Earth. It is about hills and lowlands, seas and rivers, forests and grasslands. It is also to do with why some places are warm and wet and other places are cold and dry.

geology

Geology is how you find out about the **rocks** of our world. A geologist is someone who studies how the continents, mountains, valleys and oceans were formed. A geologist uses tools to collect and test pieces of rock, and a notebook to record any useful finds.

goggles

notebook

NOTE BOOK

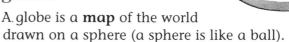

global warming

Our globe, called Earth, has had times of global cooling, as in the ice ages. Now there is a time of global warming. Gases from fuel burned in homes, cars and factories are making the globe warm up faster. See also **greenhouse effect, ice age.**

globe

A globe is a **map** of the world drawn on a sphere (a sphere is like a ball).

geyser

A geyser is a spout of hot water and steam which shoots up out of the earth. Geysers are found near **volcanoes**. They often erupt at regular intervals. The name comes from 'Geysir' in Iceland, which spouts over 50 metres high into the air.

goods

Goods are things people buy and sell. For example, cars, computers, loaves of bread and potatoes are all goods.

gorge

A gorge is a narrow, deep **valley** with very steep sides. Gorges are often cut into solid rock by rivers. There is often a **waterfall** where a river falls into the upper part of a gorge.

glacier

A glacier is like a river of ice. Glaciers are found in high, snowy mountains. The snow piles up in hollows on the mountains and becomes ice, and then moves slowly down towards low ground. Glaciers move at about one metre a day.

This glacier is in a mountain valley in Alberta, Canada.

gradient

The gradient of a slope is how steep. It is usually measured as a percentage (%) or as a ratio (one in ten).

granite

Granite is a very hard type of rock. It is usually light brown in colour, and is made of white, black and pink crystals. It forms moorlands and great cliffs. It is often used for building things like sea walls which need to be really strong.

grassland

Grassland covers wide areas in different continents of the world. The vast grasslands at the centre of North America are called the prairies. The dry grasslands of the central part of Asia are called the steppes.

21

green belt

The 'green belt' is the name for the ring of green fields around some cities. There are laws which stop owners from building houses or factories on this land. The idea is to keep the cities from growing out over all the land.

If the city of London did not have a green belt it would be more than 100 kilometres across.

greenhouse effect

The 'greenhouse effect' is how the **Earth** keeps warm. This is because of the gases in our atmosphere. Our planet is warmed up by the sun. Most of the heat stays in the air around us but a small amount escapes back into space. This keeps the Earth at a steady temperature.

Recently we have found that gases from car exhausts, electricity power stations and fires keep more of the Earth's heat in, so that less escapes into space. We think that the air around us may be overheating. If this is so, the weather and climates of the world may change.

Overheating might cause ice on Greenland and Antarctica to melt and cause low-lying countries like Bangladesh and the Netherlands to be flooded.

heat from the Sun

heat lost from Earth

heat sent back to Earth because of greenhouse gases, such as carbon dioxide

greenhouse gases, such as carbon dioxide, come from burning fossil fuels

Greenwich Meridian

The Greenwich Meridian is the most important line of **longitude** on a map of the world. It goes from the North Pole, through Greenwich in London, and on due south to the South Pole. It is also called the Prime Meridian because all other lines of longitude are measured from it.

Greenwich was chosen for this important line because at that time (in 1884) London was the main centre for exploring the world.

grid

A grid is a set of lines on a **map**. It makes it easier to find the places you are looking for.

Each square on the map's grid has a grid reference. This is a number to say how far north it is and a number or letter to say how far east it is. By looking at the number and the letter of the square, you can find it on the map.
See also **co-ordinates**.

gulf

A gulf is a great bay with a narrow mouth. 'The Gulf' is in the south-western part of Asia, between Saudi Arabia and Iran. Along its shores are the world's richest oilfields.
See also **bay**.

In the 'greenhouse effect' gases in the air are like the glass in a greenhouse. They let in plenty of heat but they do not let much heat out.

Gulf Stream

The Gulf Stream is a warm ocean **current**. It flows from the Gulf of Mexico and crosses the North Atlantic Ocean. Its waters flow as far as Norway in northern Europe. It keeps Norway's harbours free of ice, even in winter.

△

The habitat of the emperor penguin is the coast of Antarctica.

gully

A gully is a small **valley** cut into soil. It is made by water running off land where trees have been cut down.

habitat

The habitat of a plant or animal is its natural home, the place where it likes to live and has a good chance of surviving.

hail

Hail is **ice** which falls from the sky. Ice is formed in thick, dark clouds and it grows into hailstones as it is blown about in the air. The hailstones, which are usually about the size of a pea, fall in hail showers. The biggest hailstone ever found was almost 20 centimetres across.

hamlet

A hamlet is a small group of houses in the countryside which are near each other. There are usually no shops, places for religion, petrol stations or public bars. You find hamlets in farming areas a long way from towns.

harbour

A harbour is an **inlet** of the sea which gives ships shelter from the wind. Harbours usually have a dock where ships tie up. Harbours that are used by lots of ships become **ports**.

▽ *This is a harbour for small ships and boats in Villefranche on the Côte d'Azur in France.*

hardwood

Hardwood is wood from trees which grow in warm, wet areas. It is strong and heavy. We use temperate hardwoods, like oak and elm, and tropical hardwoods, like ebony and teak, for making lovely polished tables and desks.

Most temperate hardwood forests have already been cut down. Now the tropical hardwood forests are being cut down. By 2050 there may be no natural hardwood forests left. See also **rain forest, softwood.**

heath

Heath is lowland covered with tough plants. A large belt of heath stretches from the Netherlands, across northern Germany, and into Poland. Seeing heathland usually tells you that the ground is not very good for crops.

△ *Bracken is a heathland plant.*

heat-wave

When the weather is unusually hot for a few days we call it a heat-wave.

haze

Haze is air which is hard for us to see through. Sometimes smoke or dust in the air makes it hazy. On hot days the air seems to shake and we say there is a 'heat haze'. On hazy days you cannot see things two or more kilometres away.

▷ *The equator divides the Earth into the northern and southern hemispheres.*

northern hemisphere

equator

southern hemisphere

hectare

A hectare is an area of land 100 metres long by 100 metres wide. It is about the same size as two soccer pitches. Farmers today measure their fields in hectares. Small fields are less than one hectare in size, big fields are over ten hectares.

hemisphere

A hemisphere is a half of a sphere. The **equator** divides our world into two hemispheres. The northern hemisphere lies north of the equator and has continents like Europe and Asia in it. The southern hemisphere lies to the south of the equator and has Australia, some of Africa and Antarctica in it.

headland

A headland is a part of a coast which juts out into the sea. Every **bay** has two headlands, one on each side.

hill

A hill is land which is higher than other land nearby. Hills are lower than mountains and do not take so much effort to climb.

horizon

The horizon is the line that you see where the land or sea seems to meet the sky.

horn

A horn is a sharp peak on a mountain. An example is the Matterhorn in the Swiss Alps. Horns are formed where **frost** and **ice** cut mountains into these shapes.

humans

Humans are the children, women and men who make up the world's people.

Human geography is the study of where people live and how they make use of their lands. It includes the study of towns and cities. See also **geography**.

▷

The Matterhorn mountain is famous for its sharp peak, called a horn.

▽ *The centre of a hurricane is called the 'eye'.*

humus

Humus is the black material that dead leaves and grass turn into when they are left to rot on the ground. Humus becomes part of the soil and is very good for other plants to grow in. Another name for humus is leaf mould. See also **soil**.

hurricane

A hurricane is a great storm found in the tropical parts of the Atlantic Ocean. Hurricanes are over 200 kilometres across and bring violent winds of up to 200 kilometres per hour and very heavy rain. They form over the sea. They cause lots of damage in coastal areas, but die out as they move inland. Hurricanes are also called cyclones and typhoons in Asia and willy willies in Australia. See also **cyclone**, **typhoon**.

hydro-electricity

Hydro-electricity is electricity made by water flowing downhill and passing through a type of engine. It is often known as HEP (hydro-electric power). The world's most powerful HEP station is at Itaipu on the Parana River in southern Brazil. See also **power station**.

I J

◁

In the freezing climate of Antarctica ice crystals nearly cover this man's face.

ice

Ice is frozen water. You find it in cold countries in winter, when puddles and lakes, and sometimes rivers and seas, freeze over.

Ice can cover a wide area like a great blanket. This is called an ice sheet. The Antarctic ice sheet is over four kilometres thick in places.

The giant ice sheets over Antarctica and Greenland are called ice-caps.

ice age

Ice ages were times when the ice sheets were much bigger than they are today.

The last ice age ended about 10 000 years ago. During an ice age 20 000 years ago, ice hundreds of metres thick lay where the cities of Toronto, Boston, London, Berlin and Moscow are today.

iceberg

An iceberg is like a mountain of ice floating in the sea. Icebergs are formed where the ice sheets of Greenland and Antarctica meet the sea. Huge lumps of ice break away from the edge of the ice sheet and float away. Some icebergs are as big as small countries.

imports

Imports are **goods** which are bought by people in one country from people in other countries.

industry

Industry is work. There are different types of industry.

An industry that takes (extracts) things from the land or sea is called an extractive industry. Examples are **farming**, **fishing**, **forestry** and **mining**. An industry where people make things, like cars, is called a manufacturing industry. An industry where people like teachers, nurses and shopkeepers give a service is called a service industry.

Nowadays many people work in the information industry. They design computer software or make television programmes to bring information to people.

inlet

An inlet is a small opening into the land, filled by the sea. Inlets get less wind than the open sea so they are good harbours.

It is hard to believe that more than eight-ninths of the ice in an iceberg lies under the water, but it is true.

International Date Line

The International Date Line is a line drawn on maps of the Pacific Ocean near 180° (degrees) **longitude**. It marks a change of day and date. On the Asian side of the line it is one day later than on the American side. If you reach the line from Asia at 10 a.m. on Monday Asian time, the day and date changes to 10 a.m. on Sunday American time when you cross to the American side.

International Date Line
180° longitude

Asian side
(10 a.m. Monday)

N

W

E

S

American side (10 a.m. Sunday)

◁ As you cross the International Date Line the day and the date changes.

key

Maps have a key. The key tells you what the **symbols** on the map mean. See also **map**.

◁ Irrigation of cereal crops in Vietnam.

irrigation

Irrigation is putting water onto land so that crops grow well. Farmers take water from rivers, lakes or wells and let it flow through ditches or pipes onto their land.

island

An island is a piece of land with water all around it. The biggest island in the world which is not a continent is Greenland in the Atlantic Ocean.

jungle

'Jungle' is a word from India meaning 'land overgrown with underwood'. It is found where a patch of **rain forest** has been cleared and then left to grow over again. It has lots of trees, creepers and wildlife.

lagoon

A lagoon is a lake just inland from the seashore. Because lagoons are near the sea, they have salt water in them

lake

A lake is a hollow in the land which is filled with water. Lakes are usually filled by freshwater from rivers. But some lakes which are surrounded by desert, like Lake Eyre in Australia, can have salty water in them.

land

Land is the solid rock and soil that we live on. Less than one-third of the world's surface is land. The rest is ocean.

landform

A landform is part of the land with a shape that is easy to see. A hill is a landform, and so is a valley.

landmark

A landmark is a part of the countryside which stands out. It may be a hill or a church with a tower which you can see from a long way off.

△ The landforms of Monument Valley, in the USA, are spectacular to look at.

landscape

The landscape is the land surface which you can see at any time. In a city the landscape will be one of buildings and roads. In the countryside it will be of fields and woodland.

landslide

A landslide is a lot of rock and soil falling or sliding very quickly down a slope. A landslide can happen when frost or rain loosens rock or soil on a steep slope.

latitude

Latitude is a measure of how far you are north or south of the equator, which is 0°(degrees) latitude. The North Pole is at latitude 90° North. The South Pole is at latitude 90° South.
See also **equator**, **longitude**.

▽ *Latitude lines are drawn from east to west on maps.*

lava

Lava is the very hot, soft molten rock which runs out of a **volcano**.
See also **magma**.

limestone

Limestone is a grey, yellow or white rock formed from dead sea shells. When tiny sea creatures die, their shells fall to the bottom of the sea. There they get mixed up with mud and become limestone rock millions of years later. When you see **cliffs** of grey limestone or white chalk (chalk is a type of limestone), you know you are looking at what was once the sea bed.

livestock

Livestock are animals being reared on **farms**, such as cows, pigs and horses.

local

Your local area is the area around where you live.

▷ *Molten magma is called lava when it pours out of a volcano.*

location

Location is where a place is. You can give the location of a place by giving its **grid** reference on a **map**. You can also say if it is near a well-known **landmark**.

▽ *Longitude lines are drawn from the North Pole to the South Pole.*

loess

Loess is grey or yellow **soil** usually found on the edge of what used to be desert. It was picked up by desert winds and blown far away. Loess soils are very good for growing crops.

longitude

Longitude is a measure of how far east or west you are. The line for longitude 0° (degrees) passes through Europe and Africa. All longitudes are measured east or west from this line.
See also **Greenwich Meridian**, **latitude**.

longshore drift

Longshore drift is the movement of sand and pebbles along a **coast**. They are moved by **waves**.

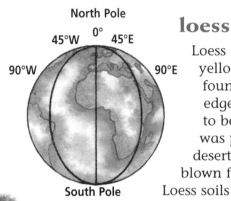
◁
The livestock kept on this farm are ostriches. They are kept for their eggs and sometimes for their tail feathers.

M

magma

Magma is the hot runny rock found under the surface of the Earth. If it escapes onto the surface it is called **lava**.

mangrove swamp

Mangrove swamps are found at some **tropical** river mouths near the sea. Mangrove trees grow in the mud in the river bed.

manufacturing

See **industry**.

map

A map is a drawing of a place as if you were looking at it from above. Maps should have **symbols**, a north arrow, a **scale** and a **key**. A large map showing special information is sometimes called a chart. For example, a **weather** chart shows the **weather forecast**.
See also **weather forecast**.

market

Market has two meanings. A market is a place or building where people get together to buy and sell things. People are also called a market when they want something. For example, the market for a Walt Disney film may be children.

market gardening

Market gardening is farming for salad crops such as lettuce and tomatoes. It is like gardening but on a much bigger scale. A lot of work is needed on the crops before they are then dug up, cleaned and taken for sale at vegetable **markets**.

marsh

A marsh is an area of low-lying wet land that is often flooded. Only moss, reeds and tough long grasses grow in a marsh. Marshes are found beside ponds, lakes, rivers and coastlines.
See also **bog**, **swamp**.

meander

See **river**.

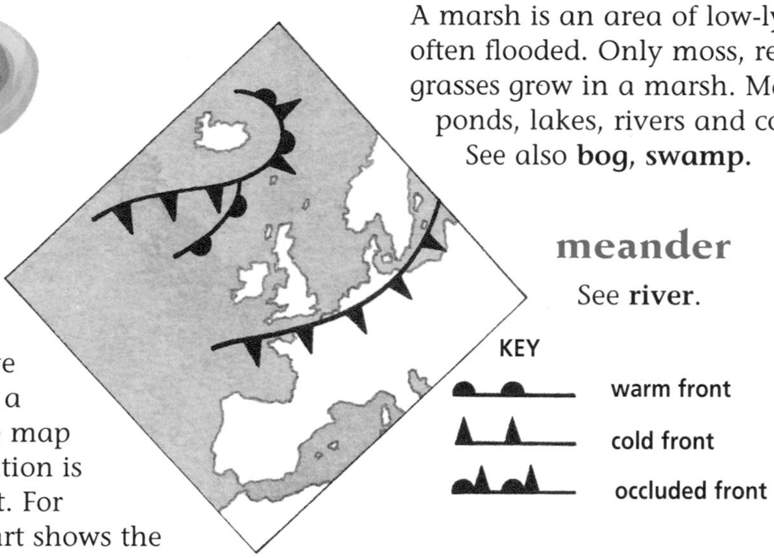

KEY

⟜ warm front

▲ cold front

▲⟜ occluded front

◁ *Here is a typical weather map, or chart, for Europe.*

migration

Migration is the movement of people from one place to another to live. People move from their home for two reasons. Some feel that they can live a better life in another place. Others fear for their lives if they stay in their home, so they move away from the danger. Animals and birds also migrate. They migrate each year at different seasons.

▽ *A floating market in Thailand where people buy and sell things from their boats.*

meridian

See **Greenwich Meridian**.

mineral

A mineral is anything found in rocks below ground which can be used by people. You may have some of these minerals in your home or school: coal, chalk, rock salt, oil, natural gas and diamonds.

mining

Mining is the work of digging holes into the ground to get **minerals** out. An example is coal-mining.

A mineral, such as amethyst, can be polished and made into jewellery.

mist

Mist is a thin **fog**. You can see further in mist than you can in fog. Officially, mist is when you can see further than 100 metres but not as far as 400 metres.

▽ *Mist is made of billions of tiny water droplets.*

monsoon

The monsoon is the rainy summer wind from the Indian Ocean which brings rain to southern Asia. When this wind blows in and it rains, people say that the monsoon has 'broken'.

moorland

Moorland is cold, windswept land where only tough plants like heather and reeds grow. It is found high on mountains or on cold lowlands near ice sheets.

motorway

A motorway is a wide **road** built for cars and lorries which travel fast. There is usually a strong barrier in the middle of a motorway. It stops cars going one way from crossing over and hitting cars going the other way. Other names for this type of road are freeway (USA), autoroute (France) and autobahn (Germany).

mountain

A mountain is a very high piece of land and is bigger than a **hill**. Mountains are often very steep and difficult to climb. The highest range of mountains on Earth are the Himalayas and the highest mountain in the Himalayas is Mt Everest. The highest mountain in the solar system is Olympus Mons, on the **planet** Mars. It is twice as high as Mt Everest.

mouth

See **river.**

nation

A nation is the people who live in a **country**. For example, the Japanese nation is all the people who live on or come from the islands of Japan.

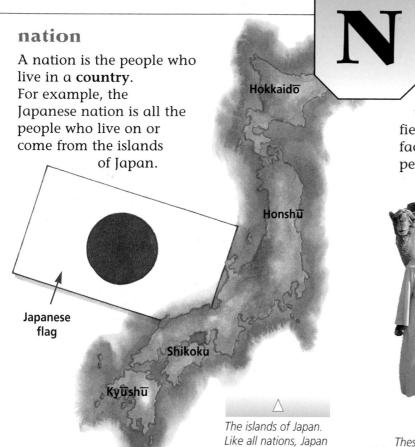

Japanese flag

The islands of Japan. Like all nations, Japan has its own flag.

Hokkaidō

Honshū

Shikoku

Kyūshū

national park

See **park**.

Nature reserves in Africa are home to many wild animals, including lions.

nature reserve

A nature reserve is an area of land where wildlife is left to live without people, in a natural way. See also **park**.

network

A network is a set of lines which cross each other as in a fishing net. **Roads** and **railways** can make a network on a **map**.

N

new town

A new town is a town which was thought about and drawn out on paper before it was built. New towns usually have things like shops, sports centres, playing fields and cycle tracks, as well as houses, factories and offices. They are often built to take people away from cities that are too crowded.

These nomads move their animals and homes in the deserts of Chad, central Africa.

nomad

Nomads are people who live and work on the move from one place to another. Most nomadic people live by keeping their sheep, goats, cattle and camels on **grasslands** on the edges of **deserts** in Africa and Asia. There are not many nomads left in the world today.

north

See **compass**.

North Pole

The North Pole is the most northerly place on Earth. All directions from the North Pole are south. It lies on the frozen Arctic Ocean, 700 kilometres from land.

oasis

An oasis is a place where water can be found in a **desert**. Oases are the only places in deserts where plants can grow and people can build homes.

ocean

An ocean is an area of water far bigger than any sea. There are five oceans in the world. They are the Arctic Ocean, the Atlantic Ocean, the Indian Ocean, the Pacific Ocean and the Southern Ocean.

oil

Oil is a slow-running liquid found in the Earth. It comes from buried plants that died millions of years ago. Oil is burned to give heat, and to power cars, ships and aeroplanes.

Oil is also very useful for making things slip over each other, so it is used to 'oil' engines.

Oil companies like BP, Mobil, Exxon and Shell are some of the biggest firms in the world.

When the iron ore has been heated, the hot metal comes out as a liquid.

oil rig

gas burning

platform

drill

platform leg

Some oil rigs are set up at sea. They are used for taking oil from beneath the sea floor.

orchard

An orchard is a small piece of land of fruit trees, such as apple and pear trees. Some large houses and **farms** have orchards. When the fruit is ripe it is picked and gathered. It may then be sold in shops and **markets**.

ore

Ore is a **rock** which has useful metals in it. When the rock known as iron ore is strongly heated, the liquid iron runs out of it. When this iron cools it can be beaten into shape to make metal things.

helicopter pad

ox-bow lake

See **river**.

ozone

Ozone is a gas in the **atmosphere**. It helps to stop harmful sunlight rays from reaching us on the surface of the Earth.

Scientists have found 'holes' in the ozone layer, and through these the rays can reach us. They think that these holes are caused by gases people let into the air from aerosols and fires. Because of these holes people should take care to cover up their skin to protect themselves from the harmful rays.

P Q

paddy field

A paddy field is a flooded field where young rice plants are grown. Rice plants need lots of water and grow well in water-logged places.

▷ *Rice is being planted in this paddy field in Madagascar.*

park

A park is an area of grass, trees and flowers in towns or cities. Everyone can walk or play in a park. National parks are much larger. They are areas of the best scenery in a **country**. People go to them to enjoy the natural scenery and wildlife. See also **nature reserve**.

pass

A pass is a low point on a ridge of mountains. If people wish to cross the ridge they usually choose to go through a pass, rather than climb over a peak.

pastoral

Pastoral land is land used for grazing animals like cattle, sheep and goats. See also **arable, farming**.

pasture

Pasture is **grassland** for animals to feed on.

peak

A peak is the highest point of a **mountain**. See also **horn**.

peat

Peat is thick, black soil found in flat, wet places. It is formed from dead plants which rot away very slowly in the water. When peat is cut into blocks and dried it can be burned on fires, like **coal**. You can buy bags of dried peat to put on the garden to help plants grow.

pebbles

Pebbles are small, rounded stones. Their smooth shape tells us that they have been rolled together by moving water, such as streams or sea **waves**.

▷

Florida is a peninsula in the south-eastern part of the United States of America.

peninsula

A peninsula is a piece of land which sticks out from the mainland so that it has water on three of its four sides.

permeable

A permeable rock is one which lets water soak into it. Common permeable rocks are **limestone** and chalk.

▽ *Large farms spray pesticides on to crops from a helicopter.*

pesticides

Pesticides are chemicals used to kill pests which eat farmers' **crops** and spread diseases. Many insects are pests.

G-BAML

HELISCOTT

petroleum

Petroleum is the 'crude' **oil** from which we make petrol for cars. It is found in **rocks** deep in the **Earth's** crust. It was formed when ancient plants rotted away in mud and sand.

place names

Most places in the world have names, and each place name has a story behind it. New York in the USA was named after York in England.

◁ These place names are well known in the city of New York, in the USA.

plain

A plain is a large, flat area of land.

plan

A plan is a kind of **map** of a small place which shows the way houses and streets are laid out. You draw plans of a place as if you are looking at the place from above.

planet

A planet is a world which moves round a star. Our own **Earth** is one of nine planets which move round our own star, the Sun. The nine planets are Mercury, Venus, Earth, Mars, Jupiter, Saturn, Uranus, Neptune and Pluto. Mercury is closest to the Sun; Pluto is furthest away.

◁ A close-up of a planet surface. The planet is Venus.

plantation

A plantation is a farm in the **Tropics** where a crop is planted and grown for sale abroad. Coffee, tea and sugar are other common plantation crops.

◁

Bananas grow quickly on plantations in the Tropics. They are shipped to Europe for sale in shops. Bananas will not grow in Europe itself.

plateau

A plateau is an area of high but flat land. The highest is the plateau of Tibet. It is over 4000 metres above **sea level**.

ploughing

Ploughing is how farmers use ploughs to cut into **grassland** and turn it over to make the soil richer. Farmers also plough farm land to make it ready for growing crops.

Sometimes farmers plough along the sides of fields rather than up and down. The little ridges of soil hold back any rainwater, so the soil is not washed away by the rain.

polar

The polar parts of the world are those areas that lie around the **North Pole** and the **South Pole**. Polar areas have a cold **climate**, with lots of **snow** and **ice**. Seven-tenths of all fresh water on Earth lies as ice in the polar ice caps of Greenland, Antarctica and the Arctic Ocean.

polder

'Polder' is a Dutch word for a type of **field** in the Netherlands. Polders are fields which used to be part of the sea-bed. The Dutch people are short of land so they built walls, called 'dykes', in shallow sea water to keep the sea back. Then they pumped out the sea.

poles

The poles of the Earth are the **North Pole** and the **South Pole**. They are the furthest points north and south on Earth.

pollution

Pollution is the harmful waste from our homes, cars and **factories** which pours into the air, rivers and seas around us. See also **waste disposal**.

Pollution can damage and kill plants and animals. It can also cause diseases in humans.

pond

A pond is a small area of water. Ponds give water-loving animals and plants a place to live.

population

Population means people. The population of a **city** or **country** is the number of people who live there. China has more people than any other country. Its population is 1200 million people.

port

A port is a place where ships stop to load and unload their **cargo**.
See also **dock**.

pot-hole

A pot-hole is a **cave** which has parts that go straight down as well as parts that go into a hillside. Most pot-holes are found in **limestone** rock. Pot-holers are people who enjoy exploring these caves.

power station

Power stations create electricity. Most do this by heating water to make steam. The steam turns engines called turbines to create the electricity. Nuclear power stations use the heat from radioactivity to boil water into steam.
See also **hydro-electricity**.

precipitation

Precipitation is another word for rainfall.
See also **rain**.

quarry

A quarry is a place where people cut **rock** out of the ground. Rock is cut or shaped to be used for building houses, factories and offices. In **limestone** quarries rock is ground up and then used for making roads and cement.

China is smaller than Canada but its population is nearly sixty times greater. Canada has 20 million people and China has 1200 million people.

railways

Railways are the metal tracks which trains run along. Trains are fast and heavy. The tracks must be laid on flat or gently sloping ground as most engines cannot pull carriages up steep slopes. So railways are often found in valleys, but not too close to rivers in case of flooding.

rain

Rain is water which falls from **clouds** in droplets. For rain to fall, the tops of the clouds must be freezing, or there must be little bits of dust or salt in the atmosphere. Ice crystals or bits of dust and salt help to form the rain droplets within a cloud.

We can measure the amount of rain that falls in one place by measuring and adding up the rainfall from each day. The amount of rain that falls each year is called annual rainfall. The wettest place on earth is Mawsynram in India. It has 11 873 millimetres of rain per year, but it only rains on 147 days. The most rainy place on earth is a mountain in Hawaii. Mount Waialeale has rain on 350 of the 365 days of the year
See also **water cycle**.

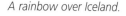

A rainbow over Iceland.

rain forest

Rain forest is found in areas near the **equator**. Rain forests are sometimes called equatorial forests. In these places it is hot and rainy all year round, and thousands of different types of trees grow quickly. **Hardwood** trees like ebony and teak grow in rain forests.

The two biggest rain forests are the Amazon rain forest in South America and the rain forest in western and central Africa.

Parrots are among the many creatures that live in the thick rain forests.

rainbow

A rainbow is an arch of colours in the sky. Rainbows appear when sunlight shines on rain and splits into seven colours: red, orange, yellow, green, blue, indigo and violet.

ranch

A ranch is a **farm** where cattle are reared on large, open spaces of **grassland**. You may have seen films of ranches with cowboys in the western USA. Many other countries of the world, such as Brazil, Argentina and Australia, have ranches too.

range

Range can mean the grasslands on a **ranch** where cattle are allowed to graze, or it can mean a line of mountains.

rapids

Rapids are the part of a **river** where its water flows very fast over rocks.

raw materials

Raw materials are things which we gather and take to factories to make something useful. From **limestone** rock we make cement for builders to make concrete. From sea water we make sea salt. From trees we get wood to make furniture.

reclamation

Reclamation means making something of little use into something that is useful. For instance, people in the Netherlands and Japan make farmland from shallow sea-beds. See also **polder**.

recycling

Recycling is using old materials again. For instance, in our towns we have 'bottle banks' to reclaim the glass from old bottles and make new glass from it.

More and more people are recycling in countries all over the world. It is a way to make the most of natural resources and things we are about to throw away. See also **waste disposal**.

reef

A reef is a line of sharp rocks on or just under the surface of the sea. Most reefs are made of **coral**.

▷ *These refugees are fleeing from Rwanda in the central part of Africa.*

refugee

A refugee is someone who has had to leave their home and go and stay somewhere else, often in another country. Refugees leave their homes because of war, famine or a natural disaster. In 1990 there were at least 15 million refugees in the world. Many people believe the problem will get worse and there will be many more refugees by the year 2000.

region

A region is part of a country or a continent and has its own character. A country may have a mountain region, a coastal region, a desert region and a region of forests.

◁ *If you take glass bottles and jars to a 'bottle bank' the glass will be recycled and used again.*

relief

Relief is how land varies in height. High relief has a lot of variation, with high **mountains** and deep **valleys**. Low relief is land which is flat or is nearly the same height.

A relief map is one of the most useful maps in geography. It shows how high the different parts of a region, country or continent are.

▷ *This is a relief map of South America showing low, high and very high land in the Andes.*

low (0–1000 m)
high (1000–2000 m)
very high (above 2000 m)

reservoir

A reservoir is a lake created by people. Reservoirs are built so that freshwater can be stored on the land until people need it for drinking, washing and making things. Cities need lots of water and every city has reservoirs nearby.
See also **dam**.

resort

A resort is a place where people go for a holiday. There are beach resorts that are popular in the summer and ski resorts which people visit in the winter.
See also **tourist**.

resources

Resources are things people can use. Natural resources are things which people find and take from the **forests**, **rocks**, **oceans** and **soil**. Gold, timber and fish are all natural resources.

We now know that we must be careful not to destroy resources so that there are plenty left for the future. In some parts of the world, such as the waters around western Europe, **fishing** is now controlled to make sure that small fish are left to grow for future fishing boats to catch.
See also **recycling**, **waste disposal**.

ria

A ria is a steep-sided river **valley** which has been flooded by the sea.

Richter Scale

The Richter Scale measures the strength of an **earthquake**. Earthquakes of below magnitude 4 are not usually felt by people. A magnitude 5 earthquake causes damage to buildings. An earthquake of magnitude 6 and over causes buildings to fall, pipes to break and fires to start.

The Richter Scale was invented by two scientists, Charles Richter (1900–1985) and Beno Gutenberg (1889–1960), in 1935.

ridge

A ridge is a **hill** which is long and thin.

rift valley

A rift **valley** is a long valley where land has sunk down between two great cracks in the land. The Great Rift Valley of eastern Africa is over 1000 metres deep and 5000 kilometres long.

1 2 3 4 5 6 7 8 9 10

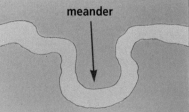

meander

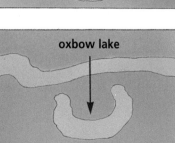

oxbow lake

Along the flat bottom of a valley, a river begins to slow down and meander. It curves left, then right, then left again, and so on. Sometimes a curve is cut off, which makes a lake. The lake is called an ox-bow lake because of its shape.

delta

oxbow lake

sea

sediment

river

A river is a large **stream** of freshwater which starts at a source and then flows through its **valley** into another river, or into the sea.

The source of a river is usually a small pool where water seeps out from a **spring** in the side of a **hill**. Sometimes a river is said to come from a **lake**.

Near their source, rivers often flow quickly in narrow steep valleys. Plunging waterfalls and white-water rapids are common. As they get nearer the sea, rivers often slow down and meander, or wind, across wide flat valleys. A meander is a winding curve or bend in a river.

Sometimes part of a meander is cut off as a river straightens its course. This makes a lake. The cut-off lake is called an ox-bow lake because it is shaped like an ox-bow – a collar for oxen.

A river flows into the sea through its 'mouth'. When the river water meets the sea it slows down and drops mud and sand which is called **sediment**. Often the river mouth is wide and shallow, with many sand bars and mud-flats, which you can see at low tide.

Smaller streams, called tributaries, flow into rivers, making the rivers bigger and bigger as they get nearer the sea.

A river basin is the area of land drained by a river and its tributaries.

The flat land near a river which floods after heavy rain is called the flood plain. Most rivers have flood plains in their lower parts where they are near the sea.

Downstream means further down the river, down the way it is flowing. A stick dropped into the water will float downstream. Upstream means further up the river.
See also **delta**.

Canoeing down a waterfall in a river can be good fun.

flood plain

meander

tributary

spring

source

rapids

waterfall

road

A road is a special surface for cars, lorries and buses to travel along from one place to another. Roads are made from tarmac, concrete or types of clay which are hard and long lasting.

rock

Rocks form the solid surface of the earth, under the soil. This solid layer of rocks is called the bedrock. Different rocks have been formed in different ways. Basalt comes from volcanoes. Limestone is formed from the seashells of dead sea creatures.
See also **clay, granite, limestone, sandstone.**

route

A route is the way of getting from one place to another. Good routes avoid steep **slopes** and places which are marshy or flooded. We prefer to plan our routes along **valleys** in low land.

This straight road is a direct route through Monument Valley, Arizona, in the USA.

run-off

When rain falls onto the land it runs off into a **stream** or **river**. Run-off is rainwater on the land before it gets into a stream.

rural

Rural means being in the countryside, away from cities. The opposite of rural is **urban**.

These are mounds of collected salt in special lagoons of salt water in Thailand.

salt

Salt is the common chemical of sodium chloride. It is found naturally in **sea** water and salt **lakes**. When salt lakes dry up, salt can be mined in the old lake bed, which is called a 'salt pan'. A 'salt marsh' is a salt-water marsh near the sea.

sand

Sand is grains (bits) of **rock** made as rocks are broken into bits by weather and water. Sand grains are bigger than dust grains but smaller than gravel. Sand can be found in **deserts** or on **coasts**. A hill of sand is called a dune.
See also **erosion.**

sandstone

Sandstone is a **rock** made up of grains of **sand** which have been forced together. Sandstone rocks are found in places where there were seaside **beaches** thousands of years ago.

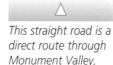

There are hundreds of satellites in space travelling around the Earth as you read this.

satellite

A satellite is something that moves around something else, like the **Earth** around the Sun. We send machines called satellites into orbit. They travel around the Earth and help us to communicate with one another.

savanna

Savannas are wide areas of land covered with grass, but with few trees. They are found in **tropical** areas of the world, like eastern areas of Africa and South America, which have wet and dry **seasons**. Savanna animals include antelope, giraffes, lions and zebras.

scale

A scale is what we use to measure distances on a **map**. If two maps as big as this page have different scales, one will show a bigger area than the other one.

scenery

Scenery is what we see when we look at a view. People often go on holiday to enjoy the scenery of another place.

scree

Scree is the pile of broken rock which lies at the floor of a mountain **cliff**. The weather widens small cracks in the cliff so that bits of rock break off and fall down the slope. On mountain slopes, the piles of scree stay for a long time. Under coastal cliffs, the sea washes away the bits of fallen rock, so it does not pile up as scree.

sea

The sea is the water which washes the shores of the land. It tastes salty, because it contains mineral salts from the soil. Most seas are the parts of the **oceans** near the land, but there are some inland seas, like the Dead Sea and the Caspian Sea in western Asia.

sea level

The sea level is the level that the sea would be if it lay flat and still. Sea level lies between the levels of high **tide** and low tide. The world's sea level is rising very slowly and low-lying countries, like Bangladesh, may become completely flooded in your lifetime.

season

A season is a time of the year when you can expect a special type of **weather**.

Near the **equator** it is nearly always warm, but there are rainy seasons and dry seasons.

Further north and south there are four different seasons: spring, summer, autumn and winter.

This tree is ① leafless in winter (but covered in snow), ② showing leaf buds in spring, ③ leafy in summer and ④ shedding its leaves in autumn.

sediment

Sediment is the gravel, **clay**, **sand** and silt (mud) carried along by a river. It is made up of bits of rock brought from upstream. When rivers are carrying lots of sediment they often look dirty or brown. The sediment is dropped when the rivers slow down. When a river floods, silt is dropped on the flat land nearby. See also **deposition**.

services

See **industry**.

 *The inside of a shopping complex in Malaysia.*

settlement

A settlement is a place where people live. Villages, towns and cities are settlements.

sewage

Sewage is solid waste that comes from houses, streets and factories. It is flushed into sewage pipes, and then drains away to sewage farms or the sea.

shanty town

A shanty town is the name for an area of a city where people have built their own houses from things which other people have thrown away. Most shanty towns are found in poor countries. People move from farms to shanty towns to try to start building a life in the city. See also **slum**.

shingle

Shingle is a lot of small pebbles on a beach. The sea grinds boulders down into pebbles and then pebbles into shingle.

shop

A shop is a place where you can buy something. In most parts of the world shops are in single rooms or buildings. In richer countries small shops are closing down as bigger shops, like supermarkets, are opened.

Any place where lots of shops are found close together is called a shopping centre. Shopping centres make it easy to shop for lots of different things in just one visit.

shore

The shore is the line where a lake or sea meets the land.

silt

See **sediment**.

Beaches are found on the sea shore. They are formed by the sea, which slowly grinds rocks down into small pebbles, called shingle. Shingle is then ground into sand.

cliff

cave

boulders

shingle

sand

sea

site

The site of a place is the land on which it is built.

situation

The situation of a place is where it is in relation to other important places nearby. For instance, the city of Cape Town in South Africa is situated on the southern point of Africa, where the Atlantic Ocean meets the Indian Ocean. Cape Town's situation makes it a natural port of call for ships going around this southern point of Africa. Ships can refuel, make repairs and trade here.

sleet

Sleet is rain and snow falling together. It makes you wet and cold.

slope

Slope is how the ground rises or falls under your feet. If the land is flat, it has no slope at all. If it rises or falls a little bit, it has a gentle slope. If it is hard to walk up or down a slope, it is a steep slope. The steepest slope of all is a **cliff**.

slum

A slum is an area of very poor housing. Slums are found in some big **cities**, although people have tried hard to clear them.
See also **shanty town**.

Slums like these have no proper power or water supplies. They can be uncomfortable and unhealthy places to live.

smallholding

A smallholding is like a small farm, where a few animals and crops are grown to sell to people nearby.

smog

Smog is **fog** and smoke found together. The fog is natural. It comes from water in the air. The smoke comes from chimneys or cars and is created when people burn fuel. Smog is harmful and can make breathing and seeing difficult.

snow

Snow is crystals of **ice** floating down from clouds. It is formed when the water in clouds freezes into ice crystals which then fall through the air and settle on the ground.

The lowest level where snow lies on a mountain through winter and summer is called the snow line. Below the snow line you find rock, moorland or trees.
See also **avalanche, blizzard**.

softwood

Softwood is the **wood** from coniferous trees, such as pine and fir. Most coniferous trees are planted so that later on they can be cut down for timber. Some softwood, like pine, is polished and made into lovely golden-coloured furniture.
See also **coniferous, hardwood**.

soil

Soil is the layer of dirt which lies over rocks on the surface of our Earth. It is made up of tiny bits of rock mixed with dead plants and is spongy, with little spaces filled with air and water.

source

A source is the small place where a stream or river starts. It is usually a spring on the side of a hill.

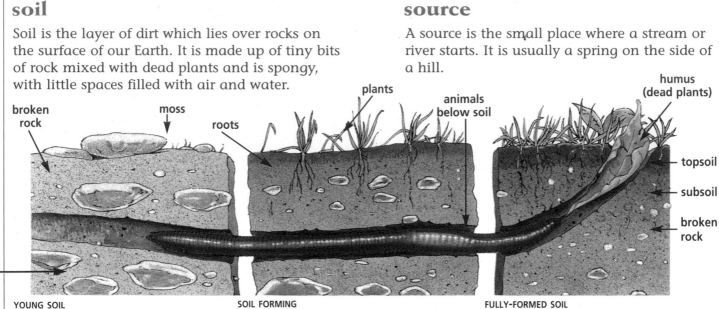

broken rock | moss | roots | plants | animals below soil | humus (dead plants) | topsoil | subsoil | broken rock | solid rock

YOUNG SOIL   SOIL FORMING   FULLY-FORMED SOIL

Crops and grass grow well in fertile soils. Loam is a fertile soil. It is a well-balanced mix of sand, clay and **humus**. Crops will not grow, or will be very poor, in infertile soils. A soil may be infertile because it does not have the minerals needed for plants to grow.

solar energy

Solar energy is energy from the Sun. It reaches us in the Sun's rays as light and heat. We can use solar energy to make electricity. Solar panels are used to collect the solar energy so that it can be turned into electricity. Solar panels are usually set up in dry places where there is little cloud to stop the Sun's rays.
See also **wind power**.

south

See **compass**.

△ *It may take thousands of years for young soil to become fully-formed soil.*

South Pole

The South Pole is the most southerly place on Earth. It lies in Antarctica. All directions from the South Pole are north.
See also **Antarctica**.

space

Space is the emptiness between the **Earth** and other places in the sky, like the Sun, Moon and stars. Space is almost empty and spaceships have to be sealed, or all their air would leak out.

spit

A spit is a low line of sand or shingle which runs from the land out across a bay into the sea. Spits are caused by **longshore drift**.

spring

Spring has two meanings. A spring is the name given to a small opening in the ground from where water under the ground comes out on to the surface. Most streams and rivers start from springs. Springs are found on the sides of hills, or on beaches, near the sea.

Spring is also the name given to the season which comes after winter.

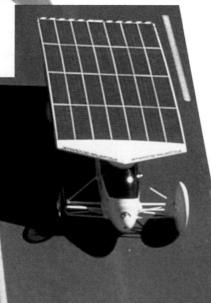

▷ *This car is filled with solar panels so that it can run on energy from the Sun.*

spur

A spur is the ridge of higher land on the side of any **valley**.

stack

A stack is a large rock in the sea with its top above **sea-level**. It is the last bit of a cliff left when the rest has worn away.
See also **erosion**.

stalactite

A stalactite is a pillar of rock which hangs down from the roof of a **limestone** cave. To remember that a stalactite hangs down, you can think that a stala*ctite* 'holds *tight* to the roof'.
See also **stalagmite**.

stalagmite

A stalagmite is a pillar of rock sticking up from the floor of a **limestone** cave. To remember that a stalagmite sticks up from the floor, you can think that stala*gmites* '*might* one day grow up to the roof'.
See also **stalactite**.

steppes

See **grassland**.

storm

A storm is a period of violent weather. Storms usually have strong winds, dark clouds and heavy rain, snow or hail.

◁ *Stalagmites and stalactites sometimes join to form a 'pillar'.*

▷

Here is a selection of symbols that may be used (as a kind of shorthand) on a map.

stream

A stream is a small flow of water which goes into a river, lake or even into the sea.

suburb

A suburb is a part of a **city** where many people go home to after work. It is mainly homes, with a few factories or offices. Most suburbs are on the outside of city centres.

summer

See **season**.

swamp

A swamp is an area of low ground that is always flooded; reeds and even trees grow up through the mud and water.
See also **bog, marsh**.

symbol

A symbol is a small, simple drawing used to stand for something, like a cross stands for a church. Symbols are used on **maps** to indicate where these things are. Symbols on maps need a **key** to tell you what they mean.

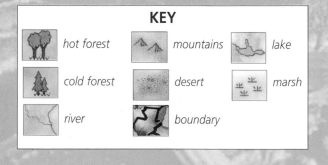

KEY

🌳 hot forest	🌲 mountains	🗺 lake	
🌲 cold forest	░ desert	✶ marsh	
🌊 river	▨ boundary		

tarn

A tarn is a small lake found high up near mountain peaks. Tarns are formed when the last ice in a **glacier** melts in the bottom of a **corrie**.

temperate

The temperate zones of the world lie between the hot lands of the **Tropics** and the cold lands near the **North** and **South Poles**. If you live in a temperate zone you will be warm in summer and cool in winter.

Third World

The Third World is an old name for the poorer countries of the world. At that time the 'First World' was the rich western countries, like the USA, the UK and Germany. The 'Second World' was Russia, the countries of Eastern Europe and China. The 'Third World' was the rest. It was mainly made up of poor tropical countries, like India. See also **developing countries**.

thunderstorm

Thunder is the loud rumbling noise you hear just after a flash of lightning. A thunderstorm is a period of heavy rain with thunder and lightning.

tide

The tide is how the sea rises and falls on the coast. The tide 'flows' in to make the high tide and 'ebbs' out to make the low tide.

High tides are caused when the Moon's pull makes the **sea level** rise. Low tides are the sea level going back to normal.

▽ Any place near the sea between 23½° (degrees) and 66½° in the northern and southern hemisphere is in a temperate zone.

66½° N

TEMPERATE ZONE

23½° N — tropic of Cancer

equator

23½° S — tropic of Capricorn

TEMPERATE ZONE

△ Temperate lands were once covered in forest. Most of these forests have been cut down and now there are fields, meadows and woods.

▽ Jericho in Jordan was one of the first towns to be built. In 6000 BC it had a wall all around it, and 2000 people lived there.

tornado

A tornado is a very violent whirlwind. Tornadoes can do a lot of damage, snapping tree trunks and flattening houses. They are most common in the states of Texas and Oklahoma in the USA where very cold air from the north meets very warm air from the south.

tourist

Tourists are people who visit another place for pleasure. Many people work in the tourist **industry**, helping other people to enjoy themselves.

town

A town is a place halfway in size between a large village and a small city.

trade

Trade is the buying and selling of **goods**.

transport

Bicycles, cars, buses and lorries, and trains, planes and boats are all types of transport. They are all ways of taking people or **goods** from one place to another.

tree line

The tree line is not really a line. It is the highest level on any mountain where trees still grow. Above this level it is too cold for trees to survive.

tributary

See **river**.

tropical

The tropical parts of the world are those areas that lie between the tropics of Cancer and Capricorn. Tropical areas have a hot **climate** all year round.

tropical rain forest

See **rain forest**.

Tropics

The Tropics are the lands which lie between the tropic of Cancer and the tropic of Capricorn. The tropics of Cancer and Capricorn are lines drawn on **maps** to the north and to the south of the **equator**. Countries in this **zone** are called 'tropical countries'.

△ In south-eastern Asia, motorcycles are a popular form of transport for the whole family.

tundra

Tundra is the name for the hardy shrubs, mosses and lichen which grow in very cold areas near the **snow** line. Tundra is found near the ice of Greenland and **Antarctica** and on the vast treeless plains of northern Canada and northern Russia.

tunnel

People dig tunnels so that **roads** and **railways** can go through **mountains** or under rivers. The Channel Tunnel goes under the sea between Britain and France.

running tunnel

passage linking tunnels

service tunnel

running tunnel

△ This train is just arriving in France. It has come from Britain through the Channel Tunnel which is 50 kilometres long.

typhoon

Typhoons are huge **storms** which sweep into eastern Asia from the Pacific Ocean. The word 'typhoon' comes from the Chinese word for 'great wind'. Typhoons are the same things as **cyclones** and **hurricanes**. They are all caused when hot seas make the air above them rise and swirl around at terrific speeds.

urban

Urban means 'to do with a town or city'. An urban bus service will run in a town, not in the countryside. The opposite of urban is **rural**.

Some animals have become urban – they have made their homes in cities and towns. This fox finds plenty of food in people's dustbins.

valley

A valley is a long line of low land between hills. Valleys usually have a river or stream running through flat, grassy meadows where the land is lowest.

Valleys have different shapes. Valleys in montain areas have steep sides with very little flat land near their rivers. In lowland areas valleys have much more flat land and the valley sides are not so steep.

Some 'dry valleys' do not have streams at all. The rainwater just soaks through into the **limestone** or **chalk** rock below.

vegetation

Vegetation is plant life, such as grass, bushes and trees.

village

A village is a small place where people live. Villages do not look the same round the world.

In a country like Britain a village has stone or brick homes, usually a church and a place where you can buy food and drink.

Villages called kraals in southern **Africa** look very different. Houses are small and made of local materials such as wood and dried mud. The people also build a fence or wall in a circle round their kraal to keep their animals in and to keep wild animals out.

volcano

A volcano is a **mountain** where hot **rock** and gas come out from inside the **Earth**. Volcanoes are found where the Earth's rock layer is thin and cracked, so that **magma**, the hot rock deep within the Earth, leaks up to the top of the mountain. Sometimes the whole volcano blows up because the hot rock and gas inside cannot get out quickly enough.

Inside the volcano there are pipes, through which the magma and gas escape. They come out of the volcano through holes

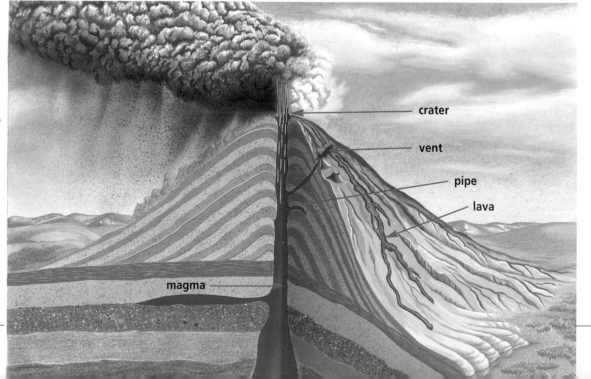

crater

vent

pipe

lava

magma

war

War is where two or more **countries** fight each other. In a civil war groups of people in the same country fight each other. In any war many people get killed or lose their homes.

War can also cause **pollution** and can damage the land. During the Gulf War, in 1991, Iraqi soldiers set fire to oil wells in Kuwait. They also caused one million tonnes of oil to spill into the waters of the Gulf.

water cooling
water freezing
heat from Sun
snow falls
rain falls
water evaporating
rivers carry water to the sea

waste disposal

Waste disposal is how we get rid of rubbish. Rubbish is what we throw away. It can be left-over food, food wrappings, worn-out clothes or broken televisions. The rubbish we throw away is taken to the rubbish tip.

Getting rid of mountains of rubbish is a problem. Some countries, like Germany, ask people to separate their rubbish into bags – one bag for glass, another for plastic, and so on. The different bags go to **factories** where the glass and plastic is recycled into new **goods**.
See also **recycling**, **resources**.

△

At this waste disposal site all the rubbish is being gathered, ready to be buried underground.

water cycle

The water cycle is the path taken by water as it goes from the sea to the land and back again. First, sea water becomes water vapour (a gas) and rises into the air. This is called **evaporation**. The water vapour drifts on the wind across to the land. Then it turns back into water droplets which form clouds. This is called **condensation**. The rain water then flows back in streams and rivers to the sea.

The Niagara River, USA, is famous for its falls, which are over 45 metres high.

waterfall

A waterfall is where a river flows over a steep drop into a pool below. Then the river flows away from the pool, now much lower down.

watershed

A watershed is a ridge of high land around a river basin. Rainwater falling within the watershed will flow into the river basin. Rainwater falling outside the watershed will flow away into other rivers.

waterspout

A waterspout is a **tornado** over the sea. A whirlwind sucks up a column of water at great speed into a thundercloud.

waterways

A waterway is a river or canal where boats carry people or **goods** from one place to another.

wave

A wave is the rolling, breaking movement of sea water or lake water onto the shore. Waves are caused by winds moving over the water.

weather

Weather is what is going on in the air above us. Fine weather has few clouds, plenty of sunshine and light winds. Stormy weather has cloud all over the sky, with strong winds and rain. Cold weather may give frost and snow. A 'heat-wave' is very hot, sunny weather.

Weathering has removed much of the rock over a period of years in Bryce Canyon, Utah in the USA.

weather forecast

It is possible to forecast the weather. This means you can say what the future weather is likely to be, although it is very hard to get it exactly right. A lot of money is spent on weather satellites and computers to try to make weather forecasting as accurate as possible, so that people and businesses can be warned if there is bad weather ahead.

The barometer is used in weather forecasting as it shows air pressure. If the pressure is falling, the weather is likely to become stormy. If the pressure is rising, the weather is probably going to become more calm.

weather front

A front is the **weather** boundary between an area of warm air and an area of cold air.

A *warm front* is a belt of weather with drizzle or light **rain**. When the rain has cleared away, the sky clears and the weather is warm.

A *cold front* is also a rainy belt of weather, but when the sky has cleared the air is colder than before.

An *occluded front* is a very rainy belt of weather. It is where a cold and warm front have caught up with each other, and have joined together.
See also **map**.

weathering

Weathering is how the weather breaks up solid rock into stones and stones into bits of dust.
See also **erosion**.

west

See **compass**.

wilderness

Wilderness is land where no one lives all the time. The biggest wilderness on **Earth** is **Antarctica**. The only people who live there are scientists.

This is a weather vane. The cockerel points in the same direction as the wind is blowing.

wind

Wind is moving air. Light winds, such as **breezes**, move slowly. Strong winds, such as gales, move quickly. Some winds are just local ones, but others travel across the Earth for hundreds of kilometres. 'Jet stream' winds are high in the air, where airliners fly. 'Trade winds' blow steadily from the east towards the **equator**. See also **Beaufort Scale**.

wind power

Wind power is the power made when the wind turns the sails on a windmill. Wind power is now seen as a 'clean' power because it does not pollute the air.
See also **solar energy**.

winter

See **season**.

wood

Wood has two meanings. First, wood is what is inside trees. When a tree is cut down it is sawn up into logs of wood. Wood pulp is what you make when you grind logs of wood into bits and mix it with water. It is soft and mushy and is squeezed into sheets and dried to make paper.

Wood also means an area of trees. A wood is smaller than a **forest**. See also **forestry**, **hardwood**, **softwood**.

world

The world is the planet **Earth** that we live on.

xerophyte

A xerophyte is any plant which can live in very dry conditions.
Cactus plants, for example, are xerophytes and can survive in deserts. They have thick, waxy skins to keep water in. They store water in their stem and have long roots to draw water up from deep in the soil.

yield

A yield is the amount of food which a **crop** gives us per **hectare**.

zone

A zone is a place on a map with a line around it. The world can be drawn with frigid (cold), tropical (warm) and temperate (mild) zones. The cold zones are near the poles, the warm zone is near the equator and the mild zones are between the warm and cold zones.

This map shows the frigid, temperate and tropical zones. The tropical zone lies between the tropics of Cancer and Capricorn.

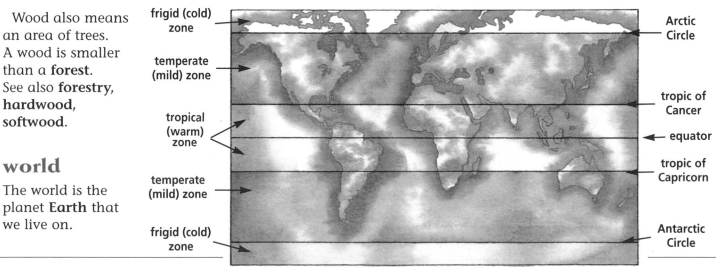

frigid (cold) zone

temperate (mild) zone

tropical (warm) zone

temperate (mild) zone

frigid (cold) zone

Arctic Circle

tropic of Cancer

equator

tropic of Capricorn

Antarctic Circle

North America

North America is made up of the countries of Canada, the United States, Greenland and Mexico, plus the seven mainland countries of Central America and the 16 island countries of the Caribbean. The west and east of the continent are highlands. In between are lowland plains, with five Great Lakes and the long Mississippi, Missouri and Mackenzie rivers.

The caribou of northern America are a prized catch by many hunters.

ALASKA
Mt McKinley
Rocky Mts

The north

The northern part of North America is cold and dark in the winter but sunny in the summer. It is one of the emptiest places in the world. Ten times more people live in the city of New York than in the thousands of square kilometres of Alaska and northern Canada. Animals such as caribou and musk oxen roam the frozen moorland landscapes.

Death Valley
Los Angeles   Grand Canyon

The centre

Travelling from the Rocky Mountains to the Atlantic Ocean you would see rich farmland, forests and large cities. The five Great Lakes along the US–Canada border and the rivers allow large ships to sail as far as the prairie grasslands in the centre of the continent. The Mackenzie River flows northwards into the cold Arctic Ocean and the Mississippi River drains southwards into the warm Gulf of Mexico.

MEXICO

Mexico City

The west

The west of North America is a place of grizzly bears and other large things. Along the shores you can see grey whales of the Pacific Ocean and on the nearby mountains grow the tallest trees on Earth, 100 metres high. Inland the high mountains and plateaux are cut by deep canyons, like the Grand Canyon, the world's deepest. Los Angeles takes up more land and Mexico City has more people than any other city in the world.

◁
Giant redwood trees are the tallest living thing in our world. They can grow to over 100 metres and live for 200 years.

▷
Grizzly bears can be very dangerous and fight with their teeth and claws as well as the well-known 'bear hug'.

ARCTIC OCEAN

GREENLAND

CANADA

> The bald eagle is the national symbol of the United States.

The east

Many of the first people to live in Canada and the United States came from Europe, and landed at what are now the old seaports of Montreal, Boston, New York and Baltimore. Then they moved west through the Appalachian Mountains, through places with lovely names like the Shenandoah Valley and the Blue Ridge Mountains. In 1969 the first people to go to the moon blasted off from Cape Canaveral, on the east coast of Florida.

Missouri R.

Great Lakes

UNITED STATES OF AMERICA

•Montreal

Ohio R.

Mississippi R.

Appalachian Mts

Shenandoah Valley

Blue Ridge Mts

•Boston

•New York

•Baltimore

>

The cactus is a desert plant that stores water in its swollen stem. It protects its water supply with prickles.

Gulf of Mexico

•Cape Canaveral

WEST INDIES

CENTRAL AMERICA

KEY

hot forest		mountains		lake	
cold forest		desert		marsh	
river		boundary			

The south

In the south lies the country of Mexico. In the west of Mexico it is dry, but all the lands next to the Gulf of Mexico are hot and wet. Tropical rain forest has grown over the ruins of the old cities of southern Mexico. The southern coast of the United States has marshy lagoons, oil rigs just off-shore and game reserves where interesting animals live, such as manatees and alligators.

FACTS
FACTS
FACTS
FACTS
FACTS
FACTS
FACTS

Area	24 million sq km	**Number of people**	450 million people
Country with biggest area	*Canada (nearly 10 million sq km)*	**Country with most people**	*USA (250 million)*
Highest point	*Mt McKinley (6194 m above sea level)*	**Lowest point**	*Death Valley (86 m below sea level)*
Longest river	*River Mississippi-Missouri (5970 km from source to mouth)*	**Biggest lake**	*Lake Superior (82 350 sq km)*

South America

There are 13 countries in the continent of South America. Brazil is by far the biggest, both in area and number of people. Argentina, Colombia and Peru are the next biggest. To the west of the continent are the snow-capped peaks of the Andes mountain range. Further east are flat grasslands and low forested plains.

The Amazon Basin

The Amazon Basin is a vast area of lowland. It contains the biggest rain forest in the world. But many of the rain forest trees are being cut down so that the land can be farmed. Some people think the area will become a desert. Today, the forest has more animals and plants than any other forest on Earth. But all of them are in danger, like the tree sloth and the anaconda water snake which are not found anywhere else.

toucan

There are a few Amazonian people who still live on the animals and plants they find around them.

The Angel Falls in Venezuela are the highest waterfalls in the world.

The Galápagos Islands

These are a group of islands about 1100 kilometres west of Ecuador. They are home to many rare plants and animals, including giant tortoises and dragon-like iguanas.

iguana

The grasslands

The new capital city of Brasília towers over the high, warm grasslands of Brazil. In the grasslands cattle are reared on huge ranches. Further south, 1500 kilometres away, the pampas grasslands of Argentina and Uruguay are like the prairies of North America. There are flat fields of wheat and, in drier parts, cattle ranches. The beef from the cattle is canned at factories in coastal cities, such as Buenos Aires and Montevideo.

KEY

dry forests

mountains

savanna

marsh

VENEZUELA

COLOMBIA

ECUADOR

PERU

equator

R Amazon

GALÁPAGOS ISLANDS

The Brazilian coast

In Rio de Janeiro, on the Brazilian coast, children love soccer. The Brazilian national side has won the World Cup four times.

This rich part of South America grows sugar and coffee and has cities like Rio de Janeiro and São Paulo which are expanding fast. Now, factories of all types are making this area more industrial. Many people live in temporary houses in the poor parts of the towns called 'favelas'.

▽

Soccer is a serious pastime in the bay of Rio de Janeiro which is overlooked by Sugar Loaf Mountain.

▽

Gauchos are cowboys who herd cattle on the ranches of the pampas grasslands.

Lake Titicaca

Lake Titicaca is 3800 metres up in the Andes Mountains, where Bolivia meets Peru. It is the highest lake in the world that ships can travel on. The local people make canoe-shaped boats out of reeds.

PACIFIC OCEAN

Brasília
Rio de Janeiro
São Paulo

BOLIVIA
Titicaca
PARAGUAY
Andes Mts
CHILE
Mt Aconcagua
ARGENTINA
Pampas
Andes Mts
URUGUAY
Montevideo
Salinas Grandes
Andes Mts
FALKLAND ISLANDS

The Andes Mountains

Three metres from wing-tip to wing-tip, the condor soars along the 7200 kilometres of the Andes Mountains, the longest mountain range on the planet. Because the Andes range stretches down from the equator to the far south of the continent, every possible type of climate and plant life is found there. Rain forest grows in the north, coniferous forest in the south, and desert, grassland and deciduous forest are in between.

▽

Llamas are native to the Andes Mountains which form the longest mountain range on Earth.

FACTS
FACTS
FACTS
FACTS
FACTS
FACTS
FACTS

Area	17.8 million sq km	
Country with biggest area	*Brazil* (8.5 million sq km)	
Highest point	Mt Aconcagua, Chile (6960 m above sea level)	
Longest river	River Amazon, Brazil (6280 km from source to mouth)	
Number of people	300 million people	
Country with most people	*Brazil* (150 million)	
Lowest point	Salinas Grandes, Peninsula Valdés, Argentina (40 m below sea level)	
Biggest lake	Lake Titicaca, Bolivia / Peru (8300 sq km)	

Africa

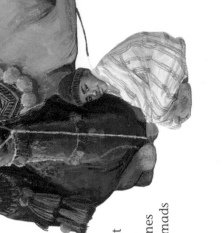

Africa is the second biggest continent (Asia is the biggest one). It is also the hottest, because it is a tropical continent. Only the very north and south of Africa are outside the tropics, so it is not surprising that the hottest place on Earth is Dalol, in the African country of Ethiopia. The two richest African countries are both mainly outside the tropics. They are Algeria and South Africa. Altogether, there are 54 countries in Africa. The biggest is Sudan, which is ninth biggest in the world, but Nigeria has the most people.

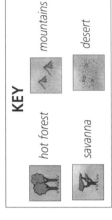

Northern Africa

Northern Africa is very dry. Only the Atlas Mountains and the coast of the countries of Morocco, Algeria and Tunisia have enough rain for crops to grow. Inland, from the Atlantic Ocean coast to the River Nile in the east, is the Sahara, the world's biggest desert. It is nearly ten million square kilometres of sand, stones and bare rock, where only nomads or oil-workers can survive.

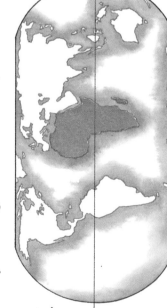

Camels are used to travel across the desert lands of northern Africa. They can also be a tourist attraction.

KEY

hot forest	mountains
savanna	desert

East Africa

Each year, millions of tourists fly to Nairobi, Kenya. They have come to see the elephants, giraffes and lions which live on the savanna grasslands of eastern Africa. As well as looking after tourists, the people of east African countries, like Kenya and Tanzania, are farmers. Some grow maize and beans, others herd cattle and sheep on the highlands near Mount Kilimanjaro and Mount Kenya. On the coast, hotels line the white sands of the beautiful Indian Ocean near the old cities of Mombasa and Dar es Salaam.

The giraffe is the tallest animal in the world and can sometimes be up to five metres tall.

Nigerian farmers grow crops such as cocoa, bananas and coconuts.

West Africa

The tall trees of the West African coast show that the land is hot and rainy. Inland it is drier and the forest gives way to tall savanna grassland. Much of the forest has been cut down and people live on farms, growing cocoa, bananas and other tree crops. Most people live near the coast in large, modern cities like Lagos in Nigeria and Accra in Ghana.

Southern Africa

Table Mountain towers over Cape Town at the southern point of South Africa. Here in the southern part of Africa there are many different types of people, each with their own way of life. This part of Africa is one of the most spectacular parts of the world. The Drakensberg (Dragon) Mountains fall away eastward in forested valleys to the Indian Ocean coast. To the west are high grasslands called 'veld', which give way to the Kalahari Desert and the Namib Desert on the dry western coast. The main city of the veld is Johannesburg, in an area where there are mines which produce most of the world's gold.

Madagascar

The ghostly lemurs, a type of ape, leap from tree to tree in the forests of this large island, off south-eastern Africa. Lemurs are only found in Madagascar, as are nine out of every ten plants and animals on the island. Over ten million people live here, farming on the grasslands and working in the rainy eastern forests.

△ *Lemurs are only found on Madagascar. This is a ring-tailed lemur.*

▽ *The life in the big cities in South Africa is very different from the traditional ways of village life. Here an African girl carries firewood to her village.*

Map labels

GHANA • Accra • Lagos
NIGERIA
CAMEROON
GABON
equator
R Congo
R Zaire (Congo)
ZAIRE
ANGOLA
ZAMBIA R Zambesi
ZIMBABWE
BOTSWANA
Kalahari Desert
NAMIBIA
Namib Desert
Orange R • Johannesburg
SOUTH AFRICA
• Cape Town
Drakensberg Mts
ETHIOPIA
KENYA
Mt Kenya • Nairobi
L Victoria • Mombasa
Mt Kilimanjaro
TANZANIA • Dar es Salaam
INDIAN OCEAN
MADAGASCAR

FACTS

Area	30.3 million sq km
Country with biggest area	*Sudan (2.5 million sq km)*
Highest point	*Mount Kilimanjaro, Tanzania (5894 m above sea level)*
Longest river	*River Nile, north-east Africa (6670 km) from source to mouth)*
Number of people	750 million people
Country with most people	*Nigeria (130 million)*
Lowest point	*Qattara Depression, Egypt (133 m below sea level)*
Biggest lake	*Lake Victoria, central Africa (69 500 sq km)*

Europe

"Europe is the jewel among continents," said a world traveller. He meant that Europe is small, lovely and valuable. There are over 500 million Europeans living in over 30 countries, each with their own proud history and traditions. The European Union is a group of European countries which work together. The colours of Europe include the dark forests of the north, the green Atlantic coasts, the sunny Mediterranean blue of the south and the white peaks of the Alps.

The fiords of Norway are long and narrow and as deep as the surrounding mountains are high.

Europe has as many traditions as other continents including national costumes and architecture.

The western part of Europe is rich in coal and oil and these are burned in power stations to provide electricity.

Northern Europe

Imagine looking out over dark, glassy water from a ferry boat coming into harbour. There are tall, snowy mountains on all sides. This is what the famous fiord coastline of Norway is like. Near the Baltic Sea the lowlands of Sweden, Finland and northern Russia have dark forests and lakes which shine blue in the long summer nights but are frozen white in winter. At midsummer, in the far north, the sun never sets. It is called 'The Land of the Midnight Sun'. In midwinter, the sun stays below the horizon all day long, so the nights are 24 hours long. Most northerners prefer the summer!

KEY

	cold forest
	marsh
	mountains

RUSSIA

R Volga

• Moscow

FINLAND

ESTONIA

LATVIA

SWEDEN

NORWAY

DENMARK

North Sea

ICELAND

BRITISH ISLES

ATLANTIC OCEAN

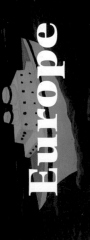

Western Europe

The British Isles, France, Germany and the Benelux countries (Belgium, Netherlands and Luxembourg) are mild and rainy. This means that farmers can grow crops and rear animals easily. As well as farming, the people of these countries have built large cities, with factories and offices. The cities are linked by modern roads and railways. This area is one of the richest places to live on our planet.

Southern Europe

Southern Europe is high, dry and hot, especially in summer. Tourists from the rest of Europe spend their holidays on the coasts of Portugal, Spain, France, Italy and Greece. They enjoy the beaches and blue seas. Inland, these countries are very hilly, with narrow roads and small farming villages. The villages are not as rich as the holiday resorts.

Central and Eastern Europe

Towering over central Europe are the mountains of the Alps. The highest peak is Mont Blanc (which means 'white mountain') reaching 4807 metres above sea level. Some alpine countries are Switzerland, Austria and Slovenia. All these countries have beautiful scenery which many tourists come to enjoy in the summer and winter. Further east from the Alps are the Carpathian Mountains which run through Slovakia, Poland and Romania, and lie in a ring around the plains of Hungary. These countries are not as wealthy as the rest of Europe, but have their own traditions and lovely cities.

The high mountains of Europe are covered in snow in the winter making them ideal for skiing and other snow sports.

The coasts of the Mediterranean countries have sandy beaches that are enjoyed by tourists from all over the world.

FACTS

Area	10.4 million sq km
Country with biggest area	western part of Russia (5 million sq km)
Highest point	Mount El'brus, Georgia (5633 m above sea level)
Longest river	River Volga, Russia (3700 km from source to mouth)
Number of people	505 million people
Countries with most people	western part of Russia (240 million) Germany (77 million)
Lowest point	Dutch polders (4 m below sea level)
Biggest lake	Lake Ladoga, Russia (18 400 sq km)

The geographical boundary between the continents of Europe and Asia cuts through the middle of some countries.

Map labels: Caspian Sea, Mt El'brus, Black Sea, BELARUS, R Dnieper, UKRAINE, Carpathian Mts, POLAND, ROMANIA, SLOVAKIA, HUNGARY, BULGARIA, CZECH REPUBLIC, AUSTRIA, CROATIA, SLOVENIA, BOSNIA, YUGOSLAVIA, GREECE, Berlin, GERMANY, Adriatic Sea, ITALY, Rome, NETHERLANDS, R Rhine, BELGIUM, SWITZERLAND, Alps, Mt Blanc, R Seine, FRANCE, Paris, London, R Loire, Mediterranean Sea, PORTUGAL, SPAIN, Madrid, R Tagus

Asia

The tiger used to be found everywhere in Asia, the biggest of all the continents. Now it is only found in the places where people do not live. Half of all the world's children, women and men live in Asia's 50 countries. Most people are crammed into the south and east of the continent. The rest seems almost empty. Asia is nearly twice as big as Africa, the next biggest continent, but is mostly within the northern hemisphere.

Pottery has been a traditional craft in Arabia for thousands of years. The pots are used for cooking and decoration. ▷

West Asia

The west of Asia is eastwards from Europe but not as far as India and China. This is why Europeans call this area the 'Middle East'. It is very dry and much of it is desert. Nomads still live in the desert, but now oil-workers are there too. They pump out most of the world's oil from under the ground. The profit from oil makes the rulers of these countries some of the world's richest people.

▽

Mount Everest has always been a challenge for mountain climbers. The Sherpas are people who live in this part of Asia and they are very skilled in mountaineering.

Central Asia

At the heart of Asia lie the cold Gobi Desert and the highest mountains in the world, the Himalayas. The region of Tibet, now part of China, is the highest land in the world. It is over 4000 metres above sea level.

Map labels: R Volga, Black Sea, TURKEY, Dead Sea, KAZAKHST, Caspian Sea, Ara, SYRIA, ISRAEL, IRAQ, IRAN, AFGHANI, JORDAN, SAUDI ARABIA, PAKIST, Arabian Sea

FACTS		
Area	44 million sq km	
Land with biggest area	eastern part of *Russia* (12 million sq km)	
Highest point	*Mount Everest*, Nepal/ China (8848 m above sea level)	
Longest river	*Chang Jiang* (Yangtze), China (5470 km from source to mouth)	
Number of people	3400 million people	
Country with most people	*China* (1200 million)	
Lowest point	*Dead Sea*, Jordan and Israel (392 m below sea level)	
Biggest lake	*Caspian Sea* (a salt lake) (440000 sq km) *Lake Baikal*, Russia (fresh water lake) (34000 sq km)	

The geographical boundary between the continents of Europe and Asia cuts through the middle of some countries

Map labels:

ARCTIC OCEAN

R Lena

R Ob

Yenisey-Angara R

RUSSIA

R Irtysh

L Baikal

MONGOLIA

Gobi Desert

NORTH KOREA

SOUTH KOREA

JAPAN

CHINA

Tibetan Plateau

Himalayas

NEPAL

Mt. Everest

BANGLADESH

INDIA

Yangtze R (Chang Jiang)

Irrawaddy R

Mekong R

South China Sea

PACIFIC OCEAN

SRI LANKA

INDIAN OCEAN

MALAYSIA

equator

INDONESIA

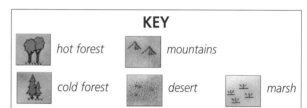

Russian, or brown, bears are very dangerous. They usually walk on 'all fours' but can walk upright using their back legs.

North Asia

The Russian bear lives in the biggest forest in the world, in the northern part of Asia. Most Asian Russians live in towns beside the few railways which cross westwards, from the Pacific Ocean to Europe. Near the Arctic Ocean coastline there is only tundra (frozen moorland). Further south is the taiga forest of fir, pine and spruce trees which can withstand the cold. Southwards again, the forest gives way to grassland. Here lies the country of Mongolia, where its few people live by herding horses and sheep.

KEY

hot forest		mountains	
cold forest	desert		marsh

Pagodas are places of worship found in Asia.

Workers harvesting rice from paddy fields in China.

East Asia

This is one of the most crowded places on Earth with 1500 million people living in farms, villages and huge cities. In wet, coastal areas, rice is the crop that most farmers grow. In the cities, people are working hard to build up their countries. The people of Japan have proved they can make things to a high standard. Now Koreans, Malaysians and Chinese people are trying to do the same.

Elephants found in Asia are smaller than those found in Africa. They have been used as working animals to carry large and heavy objects.

South Asia

Tourists come to see the elephants and lovely old temples of southern Asia. There are teeming crowds of people everywhere. India, Pakistan and Bangladesh between them have over 1000 million people, many of them living in small farms and villages. In the cities, new factories are being built. Already this region can make everything from kettles to space rockets.

Oceania

Oceania is not really one place. It is the name for a huge area which includes many groups of islands in the Pacific Ocean. It is made up of 23 countries. By far the biggest island and country in Oceania is Australia. So much of it is empty and it only has 17 million people. Only two other places in Oceania have more than a million people. They are Papua New Guinea (4 million) and New Zealand (3.5 million). Sometimes these two countries and Australia are called 'Australasia'. The people of the other 20 countries live on the hundreds of small and beautiful 'South Sea islands' of the southern Pacific Ocean.

Northern Australia

Crocodiles are found in the swampy rivers and lakes of northern Australia, where it is hot and wet. Birds of all shapes, sizes and colours fly and call to each other over the green forests of the north coast. Just off the north-east coast is the Great Barrier Reef, the largest coral reef in the world. From time to time the coasts are swept by fierce typhoon storms which blow down trees and houses.

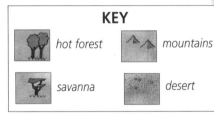

NEW GUINEA
Puncak Ja'
AUSTRALIA
• Alice Springs
▲ Uluru (Ayers Rock)
L. Eyre
• Perth
Adela

KEY

🌳	hot forest	⛰	mountains
🌿	savanna	▦	desert

◁ Kangaroos are only found in Australia. They have very strong legs and can jump up to four metres in one leap.

▽ Uluru, or Ayers Rock, rises steeply from the surrounding flat, desert-like plains.

Australia

Australia is the smallest of the world's continents. It is a very dry, sunny place, with many arid areas. Most Australians live near the south-east coast. There, the sky is sunny for most of the year but enough rain falls for crops to grow and for cities to have enough drinking and washing water. Australia is the home of animals such as wallabies, koala bears and kangaroos.

Central Australia

Uluru, also known as Ayers Rock, near the centre of Australia, is deep in the 'Outback'. The Outback is the name given to inland Australia. This area is dry desert or desert-like land with dusty red soil and low bushes. More plants grow near the waterholes, or billabongs. It is a very difficult land to live on. Only people with great skill, like the traditional native Australians, called Aborigines, are able to survive without bringing water and food in from wetter places. This is why very few Australians live in the Outback.

FACTS

Area	8.5 million sq km	**Number of people**	30 million people
Country with biggest area	*Australia* (7.7 million sq km)	**Country with most people**	*Australia* (17 million)
Highest point	*Puncak–Jaya*, New Guinea (West Irian) (5029 m above sea level)	**Lowest point**	*Lake Eyre*, Australia (12 m below sea level)
Longest river	*River Murray-Darling*, Australia (2458 km from source to mouth)	**Biggest lake**	*Lake Eyre* (9600 sq km)

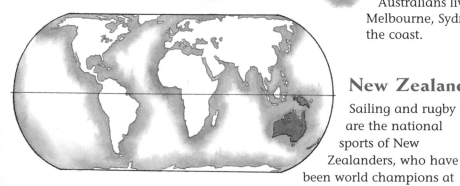

◁ *The coconut flesh and milk are used for food and other parts of the tree are used for thatching and weaving.*

PAPUA NEW GUINEA

PACIFIC
OCEAN

O C E A N I A N   I S L A N D S

Great Barrier Reef

R Darling

R Murray

• Brisbane

• Sydney

• Melbourne

TASMANIA

North Island

NEW ZEALAND

South Island

▷ *Sydney Opera House is a recognizable landmark as boats and ships enter the city's harbour.*

Oceanian Islands

Would you like to spend some time on a desert island with palm trees waving in the breeze and the sea rolling onto the silver sand? This is what many of the islands of Oceania look like. The only problem is that they are thousands of kilometres from anywhere else and there are few jobs for people. The main groups of islands are Fiji, Samoa, Tahiti, Hawaii, Kiribati, Vanuatu, Solomon, Marshall Islands and Tuvalu.

Eastern Australia

The shining white 'sails' of Sydney Opera House billow out over the city's harbour. This building was criticized when it was first built, but is now loved by Australians and is famous all over the world.

Sydney is the biggest city in Australia and Oceania, and is said to be one of the nicest to live in. It has nine months of sunshine, plenty of work for its people, beautiful beaches nearby and mountains an hour's drive away. Nine out of ten Australians live in or near the cities of Perth, Adelaide, Melbourne, Sydney and Brisbane. All these cities are on the coast.

New Zealand

▷ *Jets of hot water and steam spurting from holes in the ground are called geysers.*

Sailing and rugby are the national sports of New Zealanders, who have been world champions at both. The country has two islands. North Island is warm and sunny. It has volcanic geysers which shoot water high into the air. South Island is cooler. It has snowy mountains and fiords, like Norway in western Europe. Sheep and cattle are raised here. New Zealand lamb, wool, butter and cheese are exported all over the world.

Antarctica

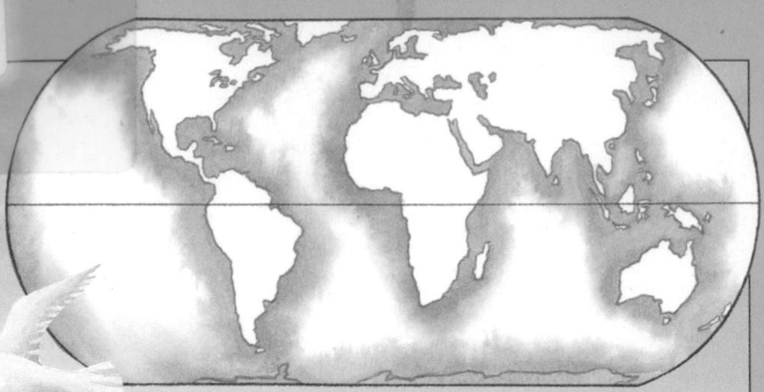

Penguins frolic on the icy coasts of Antarctica, the coldest of all the continents. It lies at the South Pole and is covered by a thick and heavy ice-cap. This ice has pushed down the land, so the land surface is 2500 metres below sea level.

Antarctica is bigger than Europe but no one makes a living there. About a thousand scientists and soldiers live in camps to study the weather, ice and rocks.

Very few types of animal can live inland on the Antarctic ice, but the coasts are rich in life with seals, whales and many types of flying birds. Recently, Antarctica was declared a protected area, so no mining can disturb the balance of nature in this great southern continent.

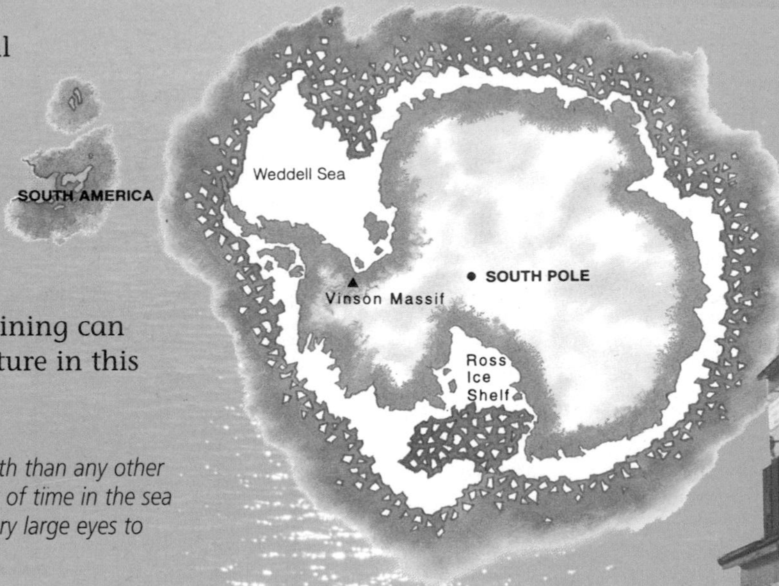

This is the sort of world map we usually see. The real shape of Antarctica is shown below.

In Antarctica the Sun sets for three months of the year from July until September when it remains dark all the time.

Weddell seals live further south than any other kind of seal. They spend a lot of time in the sea under the ice so they have very large eyes to help them see better.

Explorers and scientists have been attracted to this part of the Earth because of its natural beauty and wealth of minerals as well as its remoteness.

SOUTH AMERICA

Weddell Sea

Vinson Massif

● SOUTH POLE

Ross Ice Shelf

FACTS				
Area	13.2 million sq km		**Number of people**	1000 scientists
Country with biggest area	none		**Country with most people**	none
Highest point	*Vinson Massif* (5140 m above sea level)		**Lowest point**	2499 m below sea level (the land surface is pushed down by the weight of ice that is 4000 m thick)
Longest river	none (the land is frozen most of the year)			
Biggest lake	none (most of the land is frozen)			

64